Fertile Ground

Praise for this book

"Agroecology has been viewed as a niche intervention for small landholders for decades. It has been debated that mainstreaming agroecology is paradoxical due to the contradiction between upscaling niche innovations to produce more food in sustainable ways and the concerns for a loss of core values and principles of agroecology in the mainstreaming process. This paradoxical nature has been unbundled ever since, as examples from across the world have started emerging that it is scalable across contiguous landscapes – bringing in the community of practitioners (esp. women, small and marginal farmers, youth, and indigenous communities) in the centrality of a critical knowledge-intensive systems. This book takes you through the various facets of agroecology across the globe and brings home a key point that the universal yet contextual nature of agroecology is scalable across all spectrums."

—Swati Renduchintala, Scientist, CIFOR-ICRAF,Program Manager –
Andhra Pradesh Community-managed Natural farming programme

"Fertile Ground provides us with a global testimony of key agroecological experiences from the Americas, Africa and Europe. As interest in agroecology grows, there is a need to critically assess the evidence of its potential to fully integrate science, practice and social movements to build more sustainable food systems. This volume represents a pivotal contribution to achieving this goal."

—V. Ernesto Méndez, Ph.D., Professor of Agroecology &
Environmental Studies, University of Vermont

"If the food system is made sustainable, it'll be because of examples like those in this book. Mark Twain has some words for Big Ag: 'Few things are harder to put up with than the annoyance of a good example.' This book is filled with good examples that demonstrate the profound potential for agroecology to address everything from climate change to domestic violence. These stories aren't just inspiration -they're ammunition for a debate about the future of food and farming."

—Raj Patel, activist, academic, writer, Stuffed and Starved (2007)

"Pedro Sanchez and Dennis Garrity, two of the world's most prominent scientists and both of them ex-CEOs of ICRAF, recently agreed in an article in the Scientific American that the judicious use of plants that fertilize the soil (what are often called "green manure/cover crops") are capable of not only maintaining, but of increasing, soil fertility over the long haul. They went on to say that this fact provided tremendous hope for smallholder African farmers. This is an amazing about-face for the scientific establishment, and a major admission that agroecology can accomplish a lot more than most people had ever thought possible. I believe that agroecology will be the direction of agricultural production in the developing world. This book is one of the best books around to show us, in very practical ways, how to successfully move in that direction."

Roland Bunch, sustainable agriculture consultant, author, Two Ears of Corn:
A Guide to People-Centered Agricultural Improvement (1982)

Fertile Ground

Scaling agroecology from the ground up

Edited by
Steve Brescia

Practical
ACTION
PUBLISHING

Practical Action Publishing Ltd
25 Albert Street, Rugby, Warwickshire, CV21 2SD, UK
www.practicalactionpublishing.com

A catalogue record for this book is available from the British Library.

A catalogue record for this book has been requested from the Library of
Congress.

ISBN 978-1-78853-385-0 Paperback
ISBN 978-1-78853-386-7 Hardback
ISBN 978-1-78853-387-4 Electronic book

Citation: Steve B., (2023): *Fertile Ground: Scaling agroecology from the ground up*,
Rugby, UK: Practical Action Publishing http://doi.org/10.3362/9781788533874.

Since 1974, Practical Action Publishing has published and disseminated
books and information in support of international development work
throughout the world. Practical Action Publishing is a trading name of
Practical Action Publishing Ltd (Company Reg. No. 1159018), the wholly
owned publishing company of Practical Action. Practical Action Publishing
trades only in support of its parent charity objectives and any profits are
covenanted back to Practical Action (Charity Reg. No. 247257, Group VAT
Registration No. 880 9924 76).

Cover design by Katarzyna Markowska, Practical Action Publishing
Cover photo: Mrs. Ovoyel Edouard, a member of a peasant association in
La Victoire, Haiti, picks corn. Photo credit: Ben Depp, October 2011.
Typeset by vPrompt eServices, India

Contents

Acknowledgements

The process to create this book started with Groundswell International's global conference held in September 2014 in Haiti's Central Plateau region. We asked our international NGO partners to present their efforts to spread agroecology in transforming farming and food systems. EkoRural from Ecuador; *Partenariat pour le Développement Local (PDL)* from Haiti; Vecinos Honduras from Honduras; *Association Nourrir Sans Detruire (ANSD)* from Burkina Faso; Center for Indigenous Knowledge and Organizational Development (CIKOD) from Ghana; and Sahel Eco from Mali all participated, along with Groundswell International staff, board members, and other key resource people including Roland Bunch and Janneke Bruil.

We shared strategies, provided feedback, learned from each other, and discussed how we could strengthen this work. Two years and several additional chapters later, a short post-conference report has evolved into this book (with additional contributions from *Assessoria e Serviços a Projetos em Agricultura Alternativa (AS-PTA)* of Brazil; Steve Gliessman of the United States; and Leonardo van den Berk, Henk Kieft, and Attje Meekma of the Netherlands). The process has contributed to our own learning, and we hope will contribute to the strategies of farmers' organizations, NGOs, philanthropies, development agencies, and government actors who share the goals of creating healthier farming and food systems for people and the planet.

We would like to acknowledge, first and foremost, the many rural women and men we work with around the world, peasant farmers who are the ones leading the creation of agroecological solutions and every day constructing a better present and future. They are our inspiration and motivation, and this book is dedicated to them.

We also acknowledge the great contributions of the primary authors of each chapter. We are grateful both for their creative and tireless work in helping to generate positive changes on the ground, and for documenting and sharing them in this volume.

We would like to express particular thanks to ILIEA/AgriCultures Network, in particular Edith Van Walsum, Janneke Bruil, and Jessica Milgroom. They were key allies and thought partners throughout the process, and we benefitted greatly from their insights and edits. They also help to identify and edit the contributions from the Netherlands, as well as AS-PTA in Brazil.

Within Groundswell International, Peter Gubbels played a leading role in developing the chapters from West Africa in particular, as well as the overall structure, messages, and lessons. Cristina Hall's work in coordinating the production was vital. We would like to thank Eric Holt-Giménez and Justine

MacKesson Williams of Food First for their guidance and work in editing the book and getting it to publication.

Finally, we would like to thank the foundations and agencies who provided support to our initial conference in Haiti and in many other endeavors since: the Swift Foundation, the McKnight Foundation, the Vista Hermosa Foundation, the W.K. Kellogg Foundation, MISEREOR, and the Ansara Family Fund.

—Steve Brescia and Groundswell International

Foreword to 2023 edition

We are proud to be republishing our book, *Fertile Ground: Scaling Agroecology from the Ground Up*, as an open-access edition with Practical Action Publishing. Since our initial publication of the book in 2017, global events have only reinforced our initial motivation for documenting and disseminating these agroecological case studies and the insights and recommendations that emerge from them. We seek to honor the successes, struggles, and stories of the many people, communities, and civil society organizations that contributed to this volume, by providing it to readers at no cost.[1] Apart from this forward and a list of resource links in the appendix, we chose not to focus our time and resources on revising this edition, but rather on our ongoing work with partners in multiple countries and territories to strengthen local agroecology movements and document their achievements.

It is increasingly clear that the 'operating system' and rules by which we have organized our agricultural and food systems, and by extension our broader economic and political systems, have led us to what many are now referring to as a 'polycrisis.'[2] Any society that wishes to thrive long-term must organize itself around the goals of generating wellbeing for its citizens and maintaining the ecosystems upon which they depend. If those are indeed our goals, then both people around the world as well as our environment are shouting to us that systemic changes are required.

Agroecology is a source of positive change to transform our broken farming and food systems, which are major drivers of climate change, lead to a rising global epidemic of obesity and diet-related disease, and increase global hunger. As a field, agroecology continues to grow and generate viable and scalable solutions for a just transition toward sustainable, healthy diets, regeneration of soils and landscapes, and helping farmers and pastoralists adapt to and mitigate climate change. For those reasons it is gaining increasing support in many institutions.

In 2018, the Food and Agricultural Organization (FAO) of the UN launched a 'Scaling up Agroecology Initiative' in partnership with multiple UN agencies and governments. They see agroecology as contributing in vital ways to 15 of the 17 Sustainable Development Goals (SDGs), and they committed to "provide technical support to at least 20 countries in agroecological transition processes."[3] Many governments, international development agencies, university programmes, and funders are recognizing the importance of agroecology, while networks of philanthropists, practitioners, and academics are growing to further animate this work. Most importantly, the agroecology movement is growing among farmers, producers, and consumers of food around the world – which of course is all of us. This book, along with many other sources, seeks to contribute to this agroecological transition.

Volumes have been written about the interconnected crises we are currently facing. While the purpose of this foreword is not to catalogue these again, highlighting some recent trends brings into relief agroecology's current and potential contributions.

First, hunger is growing. In 2021, around 2.3 billion people in the world (29.3%) were moderately or severely food insecure (with 924 million people in the latter category) – or 350 million more compared to before the outbreak of the COVID-19 pandemic.[4] There is a growing debt crisis, with the percentage of low-income countries at high risk of debt distress increasing to 60% in 2022, compared to 20% a decade earlier.[5] Climate change continues to cause unprecedented patterns of drought, flooding, heat waves, and wildfires, with a frequency that has made terms like "once in 100 or 500-year event" obsolete. The COVID-19 pandemic led to over 6.9 million deaths globally by mid-2023,[6] caused the worst recession since World War II,[7] and changed how people worldwide live, work, travel, produce, and trade. Russia's invasion of Ukraine created a protracted and brutal land-war in one of Europe's bread-baskets and contributed to higher global prices and shortages of fossil fuels, fertilizers, and food. These factors, combined with fraying democracies and shifting international relations, disrupted and changed global supply chains and international trade. Finally, in early 2023 concerns about the growing and unregulated power of artificial intelligence (AI) exploded in the public consciousness as many of its leading founders and proponents themselves rang warning bells for the rest of us about its dangers.[8] In March of 2023, dozens of AI and tech leaders, led by Elon Musk's Future of Life Institute, signed an open letter calling for a six-month pause in 'giant AI experiments,'[9] asking questions such as, "Should we risk loss of control of our civilization? Such decisions must not be delegated to unelected tech leaders." They called for "work with policymakers to dramatically accelerate development of robust AI governance systems." While AI is not a central topic of this book, the issue is highly relevant because corporate actors under our neoliberal economic arrangements are seeking to use 'big data,' AI, and digital DNA to further restructure and industrialize our failing agricultural and food systems.[10]

Agroecology is, by its nature, a vast and global *human intelligence* network. It is one that is distributed across the entire world's diverse ecosystems. At its core, it is about food sovereignty, enabling localized production, control over seed, land, and trees, as well as localized markets and consumption. It is pro-people, pro-environment, pro-appropriate technological innovation, pro-local jobs, pro-ground-up economic development, and pro-democratization of knowledge, power, and resources. If a just transition away from fossil fuel dependence is necessary and feasible in the energy, transportation, and manufacturing sectors, the same is true with agriculture and food systems, which has among the highest potential for reducing greenhouse gas emissions as well as drawing down carbon out of the atmosphere. It is possible to regenerate soil fertility and food production that relies on biology and ecosystems, rather than fossil fuel-dependent chemical inputs. It is possible to transition from vulnerable, long supply chains and food miles to healthier, localized markets and circular economies.

As is stated in this book's introduction, smallholder farmers and food producers, who now represent nearly a third of the world's population and produce most of the food we eat, are a powerful force for generating ground-up solutions to our polycrisis by innovating productively with nature instead of against it. They require resources and policy support to do so. Some have commented that the scarcest resource in the world today may be imagination. We need that essential resource to see the possibilities before us and to take actions to make them realities.

The cases and lessons from this book and other documentation like it show how farmers, communities, and consumers around the world are already linking imagination to organization and to action in order to generate solutions in some of the most challenging conditions in the world. Taken together, their work points us to the principles upon which we can create healthier agricultural and food systems, as well as societies. It is our shared task to join our imaginations and actions to theirs in building a more just, equitable, and regenerative future.

Steve Brescia
Executive Director, Groundswell International
June 25, 2023

Notes

1. Electronic version is open source.
2. https://www.weforum.org/agenda/2023/03/polycrisis-adam-tooze-historian-explains/
3. https://www.fao.org/3/I9049EN/i9049en.pdf. Partners in the initiative include: The World Food Program (WFP), International Fund for Agriculture and Development (IFAD), UN Environmental Program (UNEP), Convention on Biological Diversity, World Health Organization (WHO), and UN Development Program (UNDP).
4. https://www.who.int/news/item/06-07-2022-un-report--global-hunger-numbers-rose-to-as-many-as-828-million-in-2021
5. https://oecd-development-matters.org/2022/12/07/how-to-build-global-resilience-in-a-multi-crisis-world-devat60-dialogues/
6. https://covid19.who.int/
7. https://www.brookings.edu/research/social-and-economic-impact-of-covid-19/
8. 'The Godfather of A.I.' Leaves Google and Warns of Danger Ahead, NY Times, May 1, 2023, Cade Metz.
9. https://futureoflife.org/open-letter/pause-giant-ai-experiments/
10. https://www.ipes-food.org/_img/upload/files/LongFoodMovementEN.pdf

About the authors and partner organizations

Authors

Miguel Altieri is a professor of agroecology at the University of California, Berkeley. He has been teaching at the university since 1981, and has extensive experience in scaling-up successful local sustainable agricultural initiatives in Africa, Latin America, and Asia. Currently he is advisor to the FAO-GIAHS program (Globally Important Agricultural Heritage Systems). He is the author of more than 230 publications and numerous books including *Agroecology and the Search for a Truly Sustainable Agriculture*. Miguel received a BS in Agronomy from the University of Chile and a PhD in Entomology from the University of Florida.

Cantave Jean-Baptiste is a Haitian agronomist and rural development practitioner with over 30 years of experience supporting rural development, agriculture, sustainability, and peasant organizations. He is the executive director of *Partenariat pour le Développement Local (PDL)* in Haiti. He holds a degree from the College of Agronomy and Veterinary Medicine, State University in Haiti. He speaks English, Haitian Creole, Spanish, and French.

Fatoumata Batta is Groundswell's regional coordinator for West Africa. She is one of Groundswell's co-founders and also founder of *Association Nourrir Sans Détruire* (ANSD), Groundswell's local partner in Burkina Faso. Fatou has over 30 years of experience working with rural communities. She received her master's degree in public health from Tulane University's School of Public Health and Tropical Medicine in New Orleans. She did her undergraduate work at the School of Technical Education in Paris and completed a diploma in participatory development at the Coady International Institute at St. Francis Xavier University.

Daniel Banuoku is the deputy executive director of CIKOD (the Centre for Indigenous Knowledge and Organizational Development), an NGO based in Ghana. He is a founding member of the African Coalition for Corporate Accountability in Africa, and represents Africa on the International Coordination Committee of the International People Conference on Mining. He is a member and chairman of the Environment, Agriculture and Food Security of the Lawra District Assembly. Daniel is a graduate of the University for Development Studies with specialization in Environment and Natural Resource Management. He also attended the Coady International Institute of St. Francis Xavier University in Canada.

Million Belay is the founder and director of MELCA-Ethiopia, an indigenous NGO working on agroecology, intergenerational learning, forest conservation, and livelihood improvement for local communities and indigenous peoples. He played a significant part in the establishment and activities of the Africa Biodiversity Network (ABN), and is a co-founder and coordinator of the Alliance for Food Sovereignty in Africa (AFSA). He won the National Green Hero Award for both Ethiopia and Addis Ababa in 2008. He was also a nominee to the International Forest Hero Award in 2011. He has a PhD in Education, an MSc. in Tourism and Conservation, and a BSc. in Biology.

Ross Mary Borja is the executive director of the Ecuadorian NGO EkoRural. Before taking the helm at EkoRural, Ross served as the program monitoring and evaluation (PME) Specialist for World Neighbors Andes program, where she developed a global PME system that was implemented in Ecuador and Peru. She holds a BA in economics from the *Universidad Católica del Ecuador* in Quito and a master's in professional studies from Cornell University's Community and Rural Development Program in rural sociology. Ross has authored or contributed to a number of professional works.

Tsuamba Bourgou is the executive director of *Association Nourrir Sans Détruire* (ANSD) in Burkina Faso. Engaged in rural development projects and programs since 1993, he is experienced in the strengthening of organizational capacities of farmers and their organizations in regards to project planning, management, and experience sharing. Before joining ANSD, Tsuamba facilitated the planning process and participated in the management of several programs including those of the Regional Council of the Sahel Unions programs (CRUS), the *Association Tin Tua* in the Eastern Region of Burkina Faso, and World Neighbors in Burkina Faso, Mali, and Niger. Tsuamba Bourgou studied linguistics and has a specialization in adult education.

Steve Brescia is a co-founder of Groundswell International and has been the executive director since 2009. Steve has over 30 years of experience supporting people-centered rural development, social change, and grassroots advocacy in Latin America, Africa, and Asia. He has previously worked for World Neighbors, initially supporting programs in Central America, Mexico, and Haiti and, later, on a global level; for the democratically-elected government of Haiti after the 1991 *coup d'etat* to support the restoration of constitutional democracy; and as a consultant for the InterAmerican Foundation (IAF) supporting programs in Peru, Ecuador, and Bolivia. He holds an MA in International Development from American University.

Pierre Dembélé is the executive secretary of the non-governmental organization Sahel Eco, which promotes agroecology, local economic development, and the sustainable management of natural resources in the regions of Mopti and Ségou, Mali. He is an energy engineer with over 10 years of experience in the fields of sustainable development, climate change, and energy at both the community and political levels. Pierre also coordinated the network of Malian civil society, which has been active in the field of climate change, for four years.

Edwin Escoto is the founding president of *Vecinos Honduras* and currently serves as the program and project coordinator. He is an agronomic engineer with more than 10 years of experience in rural sustainable development in both the community and regional levels across Central America including in Honduras, Guatemala, and Nicaragua. Edwin also spent four years as a coordinator with the Food and Agriculture Organization of the United Nations (FAO), where he was responsible for the technical team and all activities in the rural areas of El Paraíso, Honduras.

Drissa Gana is project coordinator at the NGO Sahel Eco in Mali. He is an agronomist with more than 20 years of experience in the fields of agroforestry, agroecology, intercommunity forest management, sustainable land and water management, and the development of value chains around non-wood forests. From 2003 to the present, he coordinated several local development projects.

Steve Gliessman is Alfred E. Heller Professor of Agroecology in UC Santa Cruz's Environmental Studies Department, where he has taught since 1981. He earned his doctorate in plant ecology at UC Santa Barbara, and was the founding director of the UCSC Agroecology Program (now the Center for Agroecology and Sustainable Food Systems). In 2008, Gliessman became the chief editor of the internationally known *Journal of Sustainable Agriculture*. He also founded and directs the Program in Community and Agroecology (PICA), an experiential living/learning program at UCSC, and co-founded the Community Agroecology Network (CAN) with his wife Robbie Jaffe. Additionally, he heads UCSC's Agroecology Research Group. He is the author of *Agroecology: The Ecology of Sustainable Food Systems,* and numerous other books and articles.

Peter Gubbels is Groundswell International's director of action learning and advocacy for West Africa, and is one of Groundswell's co-founders. He has 34 years of experience in rural development, which includes over 20 years of living and working in West Africa. Peter is the co-author of *From the Roots Up: Strengthening Organizational Capacity through Guided Self-Assessment.* Peter holds a diploma in agricultural production and management, an honor's degree in history from the University of Western Ontario, and an MA in rural development from the University of East Anglia in Great Britain.

Bernard Guri is the founder and executive director of CIKOD (the Centre for Indigenous Knowledge and Organizational Development). He has over 25 years of experience in the development sector, and also serves as the Chairperson of the Alliance for Food Sovereignty in Africa (AFSA). Bern has a BSc in agriculture from Ghana, a post-graduate diploma in rural development, and an MA in the politics of alternative development from the Institute for Development Studies in the Netherlands. He is presently pursuing a PhD from the University of Cape Coast, Ghana.

Henk Kieft has worked with NGOs in various countries and capacities including evaluation of the Dutch Fertilizer Aid to Mali, founding the Kenya

Woodfuel and Agroforestry Program, and leading the EU project "Setting up Demonstration Centres for Sustainable Agriculture and Market Study" in Bulgaria, Hungary, and Romania. He advised the Dutch Rural Network and the European Commission on the European Innovation Partnerships, as well as Closed Loop Dairy Farming programs in other Dutch provinces. Recently, Henk focused on emerging technologies based on electromagnetic influences on plants and animals, and on the need to develop intuition to manage complex farming systems.

Attje Meekma is currently chair of the territorial cooperative of the Northern Frisian Woodlands. Together with her husband and two sons, she also runs a 105-hectare dairy farm. Under their management, the outdoor grazing period of cattle has been extended, the use of antibiotics strongly reduced, and landscape and birdlife management integrated on the farm. From 2002 to 2010 she was member, and then chair, of the municipal council of Dantumadiel, where she held various posts including land policy, agriculture, and sustainability. For her work in this period, she was knighted member of the Dutch royal order of Oranje Nassau.

Pedro J. Oyarzún has extensive experience in research and rural development in Latin America and Europe. In addition to conducting research at the International Potato Center, he directed complex projects including agricultural extension, agroecological improvement, and, particularly, the strengthening of small farmers' organizations and food security. He has been an international consultant for CGIAR in Ecuador, Bolivia, and Peru. He currently works for EkoRural as an advisor on sustainable agriculture and rural livelihoods. He is the author of numerous scientific articles in international journals and has contributed to dissemination and pedagogical publications. He holds a PhD in agronomic sciences and the environment at the University of Wageningen.

Paulo Petersen is an agronomist and executive director of *Agricultura Familiar e Agroecologia* (AS-PTA), a prominent Brazilian NGO. He is also vice-president of *Aba-Agroecologia*, the Brazilian Agroecology Association, and chief editor of *Agriculturas: experiencias em agroecologia*, a magazine committed to promoting agroecological innovation processes.

Leonardo van den Berg is co-founder of *Toekomstboeren*, a peasant organization in the Netherlands that is part of *La Via Campesina*. He has also worked as an editor, journalist and researcher for ILEIA. Currently, Leonardo is pursuing a PhD at the Wageningen University. His research is on how agroecology transforms the boundaries between nature, science, and society. Leonardo also takes part in the coordinating committee of *Voedsel Anders*, the agroecology and food sovereignty platform of the Netherlands, and is a coordinator of the Dutch delegation of the European Nyeleni food sovereignty forum.

Partner Organizations

AS-PTA: aspta.org.br

Association Nourrir Sans Détruire (ANSD): www.
groundswellinternation-al.org/where-we-work/
burkina-faso

Community Agroecology Network (CAN): canunite.org

Centre for Indigenous Knowledge and Organizational
Development (CIKOD): www.cikodgh.org

EkoRural: ekorural.org

Groundswell International:
www.groundswellinternational.org

ILEIA: www.ileia.org

Northern Frisian Woodlands:
www.noardlikefryskewalden.nl

Partenariat pour le Développement Local (PDL): www.
groundswellinternational.org/where-we-work/haiti

Sahel Eco: www.sahel.org.uk/mali.html
University of California at Berkeley
Agroecology Department: food.berkeley.edu/
food-and-agriculture-related-programs-at-uc-berkeley
University of California at Santa Cruz Agroecology
Department: casfs.ucsc.edu

Vecinos Honduras: www.vecinoshonduras.org

Wageningen University: www.wur.nl/en/wageningen-
university.htm

Preface

Miguel A. Altieri, Professor of Agroecology, University of California, Berkeley, Sociedad Científica Latino Americana de Agroecología (SOCLA)

This book is a testimony to the worldwide growth of agroecology. The experiences shared here demonstrate that agroecology's intrinsic principles—used to design diversified, resilient, and productive farming systems—are strongly rooted in both science and the knowledge and practice of smallholder farmers. But the book goes beyond a simple cataloguing of techniques; it transcends technological approaches by putting agroecology at the heart of progressive social movements. It highlights how these movements are using agroecology to forge new pathways for food sovereignty, local autonomy, and community control of land, water, and agrobiodiversity.

This is important because agroecology is sometimes removed from its political context and defined solely as a science, a practice of applying ecological principles to the design and management of sustainable farms. This simplification invites a variety of competing narratives—such as integrated pest management, organic farming, conservation agriculture, regenerative agriculture, ecological intensification, and climate-smart agriculture—all of which structurally de-center agroecology and ultimately offer only minor adjustments to industrial farming.

For many agroecologists, including the authors of this book, the very systems that traditional farmers have developed over centuries are a starting point in the development of new agricultural systems. Such complex farming systems, adapted to local conditions, have helped smallholders farm sustainably in harsh environments, meeting their subsistence needs without depending on mechanization, chemical fertilizers, pesticides, or other modern agricultural technologies. Guided by an intricate knowledge of nature, traditional farmers have nurtured biologically and genetically diverse farming systems with robustness and built-in resilience. These traits are essential if agriculture is going to adapt to a rapidly changing climate, pests, and diseases. Just as importantly, it helps smallholders cope with volatile global markets, technological monopolization, and corporate concentration.

A salient feature of traditional farming systems is their high level of bio-diversity, which is deployed in the form of polycultures, agroforestry, and other complex farming systems. Guided by an acute observation of nature, many traditional farmers have intuitively mimicked the structure of natural systems with their cropping arrangements. In agroecology, examples of such "biomimicry" abound. Studies of smallholder farming systems show that, across biophysical and socio-economic conditions, there is a broad range of biodiverse farming systems (intercropping, agroforestry, crop livestock integrated systems, etc.) that sustain a series of important ecosystem services— such as pest regulation, soil health, and water conservation—and enhance both productivity and climate resiliency. Farmers don't simply add companion species at random; most associations have been tested for decades, if not centuries. Farmers maintained them because they strike a balance between farm-level productivity, resilience, agroecosystem health, and livelihoods.

Modern agroecosystems require systemic change and new, redesigned farming systems will emerge only through the application of well-defined agroecological principles. These principles can be applied by way of various practices and strategies, and each will have different effects on productivity, stability, and resiliency within the farm system. Agroecological management leads to optimal recycling of nutrients and organic matter turnover, closed energy flows, water and soil conservation, and balanced pest-natural enemy populations, all key processes in maintaining the agroecosystem's productivity and self-sustaining capacity.

The challenge to align modern agricultural systems with ecological principles is immense, especially in the current context of agricultural development, in which specialization, short-term productivity, and economic efficiency are the driving forces. By highlighting locally grounded examples of agroecological reclamation and innovation, *Fertile Ground* provides evidence that successful alternatives are attainable.

Preface

Million Belay, Director, Alliance for Food Sovereignty in Africa

For centuries, Africa has been the battleground for interests, initiatives, and ideas coming from the Global North. The "New Green Revolution" currently pushed by international companies to transform African agriculture into a high-input, industrial model, could be the most devastating of all. The agribusiness industry, powerful western governments and lobbyists, and philanthrocapitalists, aided by aggressive academicians and poorly informed country-level bureaucrats, have created a powerful but simple development narrative of "science" and "technology."

It goes something like this: "Despite progress, one in four Africans is hungry, and every one African child in three is stunted. Food demand will rise by at least 20 percent globally over the next 15 years with the largest increases projected in Sub-Saharan Africa. Agro-industrial technologies are the solution. The key need is to promote them with farmers across the continent."[1]

I once participated in a meeting attended by a senior official from the Common Market for East and Southern Africa (COMESA). He said, "In Africa the sale of certified seeds accounts for only 5 percent of farming, while in Europe the figure is over 80 percent. Europe is food sufficient while Africa is not, and seed trade plays a huge part." His statement was intended to raise a sense of incredulousness in African governments, and to promote corporate-benefitting seed laws and the commercialization of seeds as a precursor for the transformation of agriculture. I asked him, "Did you account for the trade in seeds carried out in the tens of thousands of rural markets? Is that not seed trade? Does it have to be traded by a company to be considered a trade?" I wanted to point out that seed improvement and trade is alive and active in Africa, and has been for many centuries. He had no answer.

With the stage set by this narrative of African under-development, the recommendations for market-led agriculture with more agrochemicals, hybrid and genetically modified seeds, and a highly complicated set of ideas for managing information related to agriculture, soon follow. The effort is to substitute African farmers' knowledge about the food system with new, commercial forms of knowing.

In a meeting recently concluded at the time of writing and organized by the African Green Revolution Forum (AGRF), which is led by the Alliance for a Green Revolution in Africa (AGRA), companies, philanthrocapitalists, and banks pledged US$70 billion to transform African agriculture. Transform it into what? Why all this investment of money? The simplistic answer: to transform African agriculture into a business-friendly sector for foreign companies. Yet why are agrochemical and seed companies' interests, as opposed to those of African farmers, being put at the center of the solution?

The Alliance for Food Sovereignty in Africa (AFSA) is a network of 25 networks of African farmer organizations, NGOs, and consumer groups working in 40 countries. We represent literally hundreds of practical initiatives across the continent to strengthen and advocate for African agriculture. We seek another kind of transformation: one for *food sovereignty* and the agricultural concept of *agroecology*.

Quite a number of African countries have gone in the direction of industrial or high-input agriculture. Southern and Western Africa, where many farmers rely on hybrid seeds and agrochemicals to produce food, are hit the hardest. Once engaged in this model, it is a challenge for many farmers here to transition to more agroecological practices. Their soil is already hooked on agrochemicals; their seeds will not work without fertilizer; and the system is very water-hungry. As a result, the farmers are deeply in debt. While it is relatively easy to begin applying the technology packages of industrial agriculture—and no one is advocating to make farmers' lives more difficult—this ease is short lived. Many farmers are complaining about the increasing price of seeds and fertilizers, the death of their soil due to the application of fertilizers, and heavy water demands. If there is drought, the result is a poor harvest—followed by deepening debt. In many cases, farmers lose ownership of both seed and land.

To demonstrate that agroecological knowledge is the future of addressing population growth, food insecurity, cultural erosion, urbanization, environmental degradation, and even climate change, the AFSA has collected success stories from agroecological farms from all over Africa. The case studies, like those in this volume, tell a very interesting story: agroecology increases productivity as it improves soil. It increases farmers' income because agroecological farmers do not have to spend their money buying agrochemicals. Crop diversity is high, which lowers risk and enhances resilience. It also improves nutrition and health. Unlike industrial agriculture, agroecology integrates livestock into the farming mix, increasing nutrition and shielding farmers during crop failures. Manure is ploughed back into the soil, improving production. More than anything, the case studies prove that we can fulfill the function of farming, which is feeding people and supporting the wellbeing of both the people and the earth, by working with, instead of substituting for, farmers' knowledge.

This book, *Fertile Ground*, highlights vital strategies to make agroecology a widespread reality in Africa and beyond, and to save ourselves and our planet. We need to support indigenous knowledge and innovation. We need to integrate agroecology into all policy arenas. We need to raise the consciousness of consumers about the sources of their food and the importance of healthy and nutritious food. We need to incorporate it in all levels of education so that we do not produce an army of misinformed academics who push industrial agriculture down our throats. Agroecology is not against science and technology, yet we need to facilitate the innovation of technologies that really improve the lives of farmers in lasting ways. We need to encourage small agricultural business to engage in agroecology.

I believe change is coming. Academic institutions, governments, faithbased organizations, and business are talking about and promoting agroecology. UN bodies are opening up to the concept and the practice. World social movements are more and more organizing under the umbrella of agroecology.

Agriculture is the single most powerful force unleashed on our planet since the end of the ice age. If we do not do something about agriculture's current impact on our planet, we are in deep trouble. The bright side, as the collection of cases in this volume show, is that we are making agroecology our story, our solution, and our future.

Note

1. As described in: The World Bank. "Boosting African Agriculture: New AGRA-World Bank Agreement to Support Farming-Led Transformation." Press Release, April 20, 2016.

INTRODUCTION
Pathways from the crisis to solutions

Steve Brescia

Challenges we face

Most people think that we can end hunger by increasing global food production. Agricultural modernization is often framed as the pathway. In reality, the world currently produces enough food to feed 10 billion people; that's more than enough for the seven billion that currently inhabit the planet, and the nine billion projected by 2050 (FAO, 2009). Despite a surplus in the *quantity* of production, most official estimates find that nearly 800 million people are hungry, and even more—approximately two billion—are malnourished (FAO et al, 2015). Tragically, the great majority of the hungry are themselves small-scale farmers and farm workers in the Global South.

Worse yet, these estimates of hunger may actually be too low. While the Food and Agriculture Organization of the United Nations (FAO) estimates that people are hungry only when their caloric intake falls below the "energy requirements for minimum activity levels" (which range by country from 1,651 calories/day to 1,900 calories/day), most poor people do not lead sedentary lives and are engaged in strenuous physical labor, meaning they require much more than the FAO's minimum level. For example, rickshaw drivers in India typically require 3,000–4,000 calories per day. Research shows that if hunger were measured according to the level of calories that people actually need for "normal" activity, the number of hungry rises to 1.5 billion people. When measured according to the level of calories required for the sort of "intense" activity carried out by most family farmers and poor people in the world, the number of hungry rises to 2.5 billion (Hickel, 2016). At the same time that many people go without enough food, another 1.9 billion people are overweight or obese due to dependence on processed foods and lack of access to healthy food (WHO, 2016).

Asking who feeds the world—and with what resources—leads to more disturbing statistics. While industrialized agriculture uses 70 percent of the world's resources available for agriculture (land, water, inputs, energy, etc.), it produces only about 30 percent of the food that people consume. This is because much of the production of industrialized agriculture goes to biofuels and animal feed. Conversely, family farmers use only an estimated 30 percent of the resources, but produce over 70 percent of the food consumed in the world (ETC Group, 2013). Moreover, industrial agriculture is the major reason why agriculture contributes to at least one third of the world's greenhouse gas emissions (Gilbert, 2012).

The current global, industrial food and agricultural system is fraught with contradictions that prevent it from ensuring human wellbeing or the sustainable management of our planet's resources. We need to change it. Continuing down the path of "business as usual," or simply attempting to extend the present system of industrialized agriculture to those resource poor farmers who do not yet farm this way, will not resolve these problems.

The real costs of our food system

The real costs of our food system are further highlighted by a 2012 study on sustainability trends carried out by international consultancy KPMG (Figure I.1). The authors found that food production had the "largest external environmental cost footprint" (US$200 billion) of all the 11 sectors analyzed in the report; even mining and oil production had a smaller footprint. In fact, food production was the only sector in which the total value of externalities (costs assumed by society) exceeded total profits (KPMG International, 2012:56).

These environmental externalities (22 environmental impacts, including greenhouse gas emissions, groundwater abstraction, and waste generation (ibid:55)) do not even take into account other real costs associated with our food system, such as the loss of human potential and lives from malnutrition; the cost of humanitarian assistance to keep people alive (nearly US$2 billion a year just for nine countries alone in the Sahel between 2014 and 2016 (OCHA, 2014)); or global healthcare costs related to diseases associated with excess weight and obesity (diabetes, heart disease, cancer, etc.), estimated at US$2 trillion (Dobbs et al, 2014). Various efforts are underway to improve "true cost accounting" to make visible and measure such costs and to mainstream them into decision-making.[1] Doing so will help to illuminate the irrationalities in our current food system, and could push policy makers to support more sustainable alternatives.

Two potential pathways forward

We clearly need to pursue better alternatives. Fortunately, they already exist. Although a growing body of research and abundant farmer testimonials point to agroecology as the most productive, sustainable, and just pathway forward, policy makers continue to debate intensely between this and industrialized agriculture as competing visions for the future of our food systems. The pathway we choose will have profound implications for people and our planet.

The vision of highly industrialized agriculture is well-represented by the logic of Syngenta (Figure I.2), one of the world's largest producers of agricultural chemicals and seeds.[2] As seen in the diagram, the primary goal of the paradigm is to increase productivity, and to maximize profits among farmers and the nonfarm businesses and corporations involved in agricultural commodity chains. Most profits accrue to the latter. Subsistence smallholders

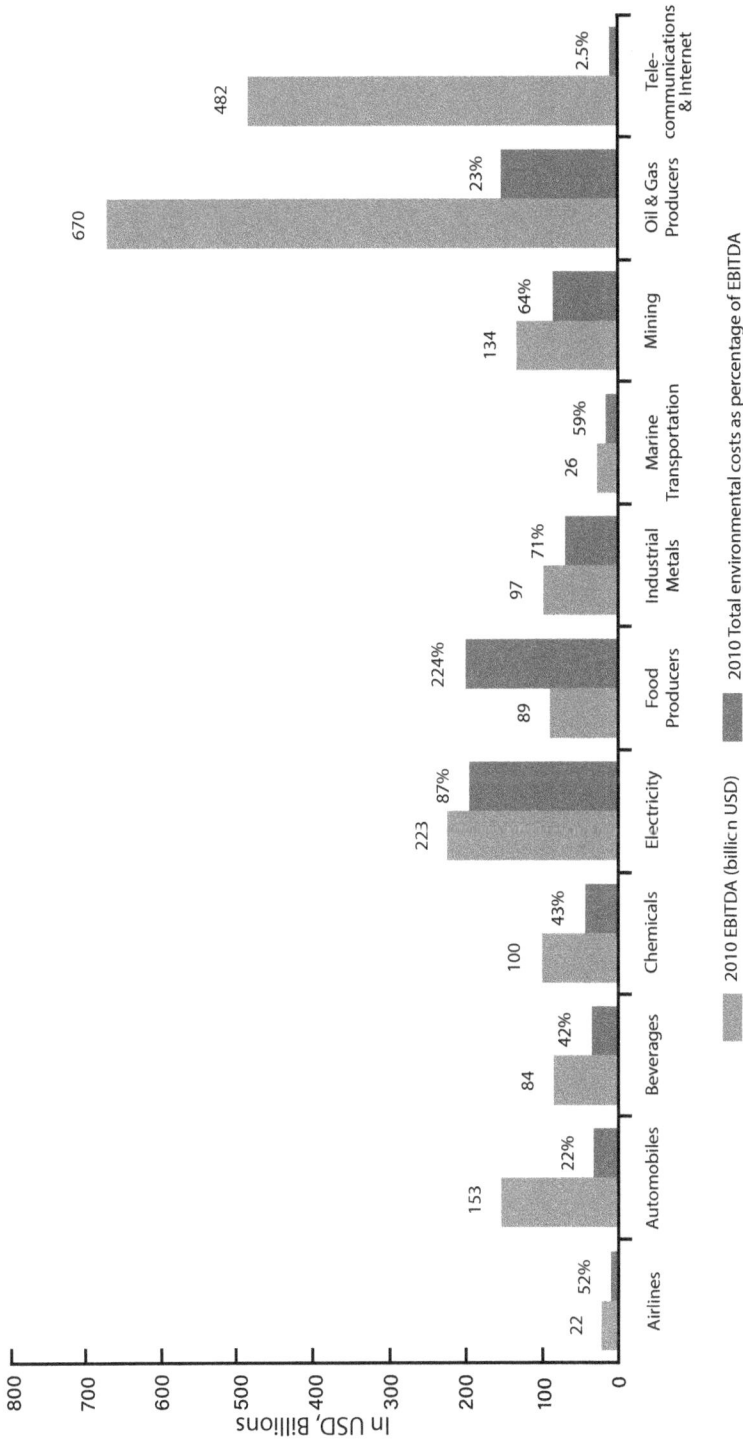

Figure I.1 The real costs of our food system.
Source: KPMG International. *"Expect the Unexpected: Building business value in a changing world."* 2012

Legend:
- 2010 EBITDA (billion USD)
- 2010 Total environmental costs as percentage of EBITDA

Data by industry:
- Airlines: 22, 52%
- Automobiles: 153, 22%
- Beverages: 84, 42%
- Chemicals: 100, 43%
- Electricity: 223, 87%
- Food Producers: 89, 224%
- Industrial Metals: 97, 71%
- Marine Transportation: 26, 59%
- Mining: 134, 64%
- Oil & Gas Producers: 670, 23%
- Telecommunications & Internet: 482, 2.5%

Y-axis: In USD, Billions (0 to 800)

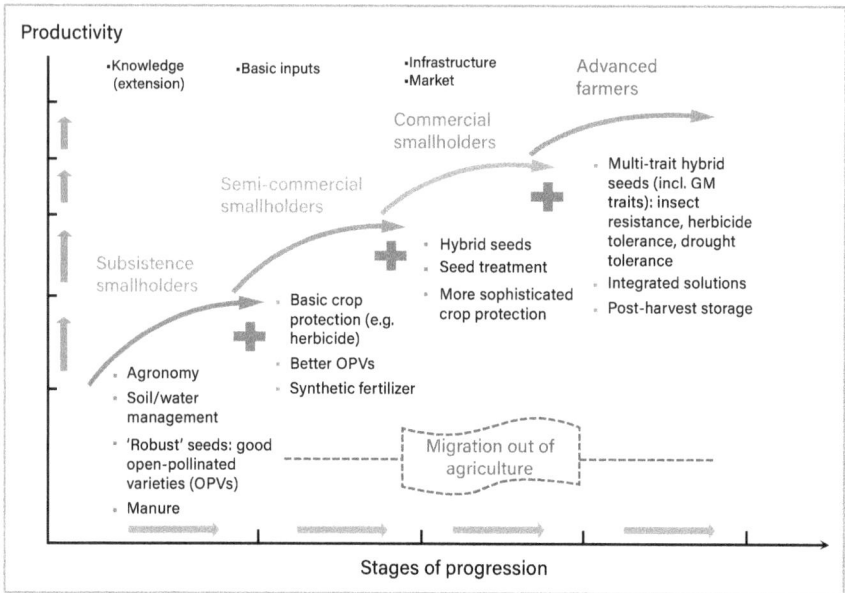

Figure I.2 The logic of Sygenta.
Source: Zhou, Yuan. "Smallholder agriculture, sustainability and the syngenta foundation." Syngenta foundation for sustainable agriculture, april 2010, p. 4

are given two options: to migrate out of agriculture or to become "advanced farmers." An advanced farmer, according to this paradigm, is one who adopts multi-trait hybrid and genetically modified seeds, chemical fertilizers, pesticides, herbicides, and other complementary practices, and that operates on a large-scale. The vision assumes that the primary role of those who remain in farming is to adopt and purchase the new technologies sold by Syngenta and other agribusinesses. It is not clear where the hundreds of millions who will be "migrated out" of farming will go, or how they will survive. According to the cases in this and other reports, experience indicates that many of these people will be deprived of land and basic resources to survive; will become more vulnerable by migrating to urban slums or across borders; or will be pushed deeper into cycles of debt, dependency, and poverty.

An alternative vision is represented in the International Assessment of Agricultural Knowledge, Science and Technology for Development (IAASTD), an international research process carried out from 2005–2007 involving 110 countries and 900 experts from around the world (Figure I.3).[3] The IAASTD process understood that agriculture is multifunctional. Researchers didn't just ask the question of how knowledge, science, and technology can maximize productivity, but rather how they can "reduce hunger and poverty, improve rural livelihoods and human health, and facilitate equitable environmentally, socially, and economically sustainable development" (IAASTD, n.d.). In response to this wider framing, the report concluded that we need to transition from our current conventional agricultural

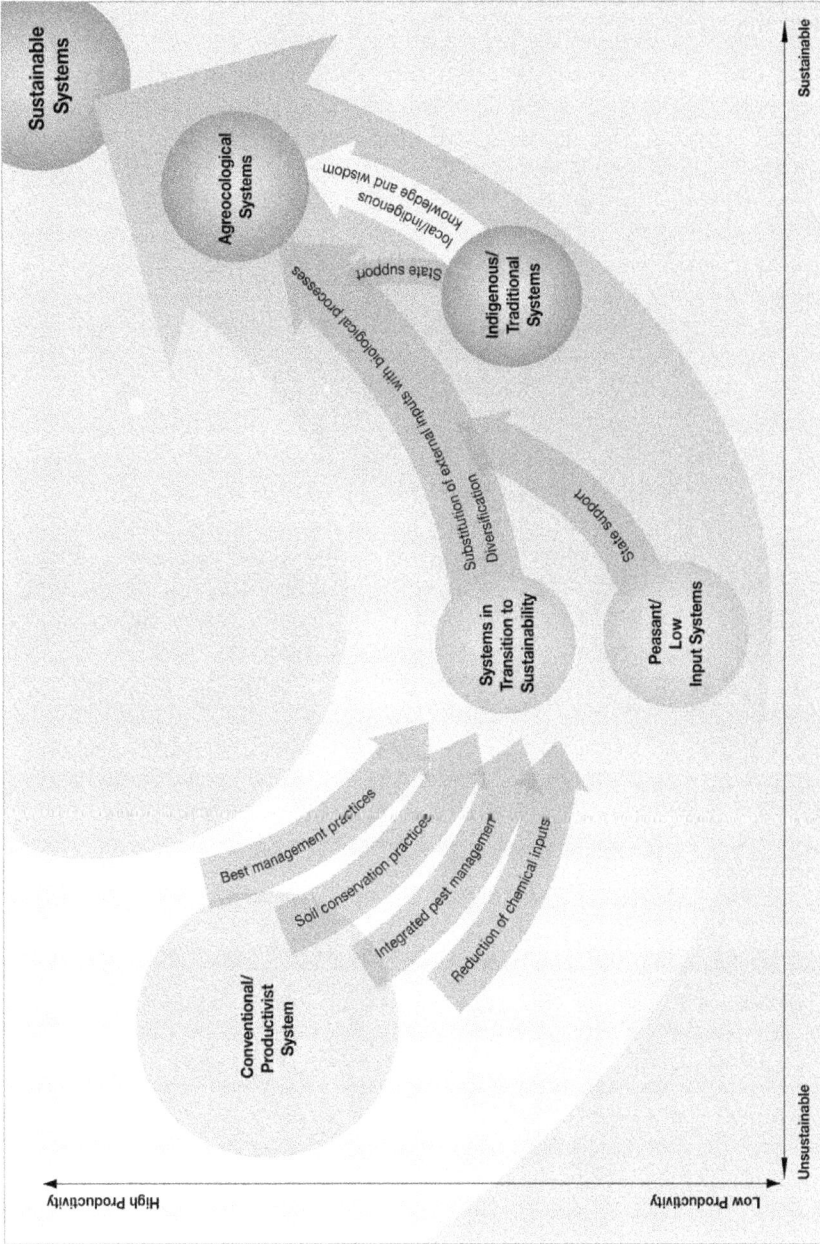

Figure I.3 Vision of the IAASTD.
Source: IAASTD. "Towards multifunctional agriculture for environmental, social and economic sustainability."

system, which privileges high productivity yet is very unsustainable, to a more agroecological one characterized by both high productivity *and* high sustainability. IAASTD also recognized that, while all farmers need to build more sustainable farming systems, their starting points and pathways vary according to context. The vision described by IAASTD and other proponents of diversified, sustainable farming also assumes an important role for family farmers as agents of change. In it, they are treated as central participants in agricultural innovation, generators of scientific knowledge and appropriate technologies, and key actors in policy-making.

The IAASTD vision already exists on the ground, in many pockets of agroecological practice. There are about 2.5 billion people in the world, on 500 million farms, involved with smallholder family agriculture and food production (FAO, 2016). Their creative capacity to farm productively and sustainably *with* nature, instead of *against* it, is perhaps the most powerful force that can be unleashed to overcome the interlinking challenges of hunger, poverty, climate change, and environmental degradation. This is the essence of agroecology.

Agroecology can be defined as: "the application of ecological concepts and principles to the design and management of sustainable agro-ecosystems" (Altieri, 1995). At its core, it is people and farmers' organizations innovating, building upon and combining local knowledge with new information, and emphasizing farms' biological processes, as opposed to external and chemical inputs.

Even before the term "agroecology" was coined by scientists, many farmers and practitioners around the world had long known that this approach to farming was an effective strategy for improving production, biodiversity, and food security; empowering communities and farmers' organizations; and sustainably managing and even regenerating natural resources. Three "streams" combine to make agroecology a growing force for positive social change today:

> **Agroecology as a Practice** – Farmers continuously innovating with nature, using principles and practices to enhance the resilience and ecological, socio-economic and cultural sustainability of farming systems
>
> **Agroecology as a Science** – The holistic study of agro-ecosystems, merging ecology with agronomy, and including human and environmental elements
>
> **Agroecology as a Movement** – A way of farming, and a process for ongoing innovation of farming systems, central to the broader social movement for food sovereignty

Food sovereignty, as articulated by the global farmers' movement *Vía Campesina*, is "the right of peoples to healthy and culturally appropriate food produced through ecologically sound and sustainable methods, and their right to define their own food and agricultural systems" (Vía Campesina, 2007).

In the cases included in this book, actors combine streams of agroecology to help "scale" the concept at different levels. The stark contrast between agroecology and the vision of industrial agriculture promoted by more economically powerful agents is further represented in Table I.1.

Table I.1 The logics of agroecology and of industrialized agriculture

	Agroecology	*Industrialized Agriculture*
Goals	To optimize benefits related to the multi-functional nature of agriculture, including production, environmental stewardship, resilience, nutrition, family and community wellbeing, and sustaining cultures	To maximize production and profits
Costs	Captures true costs, regenerates natural resources, mitigates climate change	Externalizes over US$200 billion of environmental costs annually
Sources of knowledge and innovation	Family farmers experimenting and working with nature; coordinating with scientists, government agencies and NGOs; combines traditional knowledge with modern scientific knowledge	Major agribusiness, which is increasingly concentrated, produces inputs purchased by farmers
Ways of sharing innovations (Extension Strategies)	Farmer-to-farmer learning, sharing and extension; co-creation of knowledge; farmer organizations and networks	Conventional extension and training; promotion of technology packages (seeds, chemical fertilizers, herbicides, and pesticides) through private companies or government ministries
Soil Fertility	On-farm soil conservation and soil improvement through physical (o.g., contour barriers) and biological processes (cover crops, green manures, agroforestry, composting, etc.)	Chemical fertilizers purchased from off-farm sources
Seeds	Farmer-managed seed systems and seed saving; improved selection, storage and management of local and bio-diverse seed varieties	Hybrid and genetically engineered seeds that are patented; legal constraints on farmers to save or exchange seeds; farmer dependence on purchasing from suppliers
Water management	Soil and water conservation; increased organic matter in soils to improve water holding capacity; rain water harvesting; appropriate technologies such as micro-irrigation	Larger-scale infrastructure, such as dams and irrigation
Biodiversity	Diversified farming systems (seeds, crops, livestock, fish, trees); Diverse seed varieties	Monocropping and limited seed varieties, purchased from suppliers
Markets	Emphasis first on local food needs; improved local market linkages	Emphasis on export markets, corporate value chains, biofuels, and livestock feed

Why agroecology?

Ever since people first began to domesticate grains and animals in the Fertile Crescent some 11,500 years ago, farming has been about people innovating with nature to produce food. Farmers selected, improved, and saved seed varieties; developed diverse cropping systems; conserved soil and maintained and improved its fertility; captured and used water; created tools; managed pests, plant diseases, and weeds; processed and stored food; and consumed, shared, traded, or sold their production. In these ways, cultures and agriculture have co-evolved over centuries, all around the world. The dramatic expansion of the industrialized agricultural model has largely occurred just in the last 70–100 years.

The first goal of any political and economic system should be to ensure that the people within it are fed, and that the environmental resources upon which they depend are sustained. Through history, when conditions have changed (sometimes due to human activity), farming systems and societies have had to adapt. Some have failed to do so and have collapsed. Current examples of changing conditions in this book include farmers in West Africa and Central America needing to find new ways to manage soil fertility as declining availability of land makes their traditional cycles of clearing and fallowing no longer viable; changing consumer habits in Ecuador and California; and the impacts of climate change and water availability as experienced around the world.

Smallholder farmers have the creative capacity to innovate and generate real solutions to these challenges, but they have been marginalized or exploited within our political and economic systems, their potential contributions disregarded. This book brings forward examples of family farmers acting as agents of change, rather than passive recipients and consumers of inputs. We see them contributing to the creation of healthier farming and food systems, as well as to more democratic, just, and sustainable societies.

For the authors of this book, agroecology is a people-centered and farmer-centered process of agricultural innovation and rural development. It is a process that both requires and strengthens the agency, creativity, and power of farmers and their organizations.

Why this book?

Multiple reports have already documented the benefits and techniques of agroecological farming. Others have detailed the policy recommendations required to support it (See Appendix 2: Literature on Agroecology for a list of both). In spite of this, agroecology is not nearly as widespread as it could and should be. How do we change this reality so that agroecology becomes a more prevalent part of our local and global food and agricultural systems? Building on this body of practice and research, this book addresses two key questions:

1. What strategies work to spread agroecological farming more widely?
2. How can this contribute to systemic changes in our farming and food systems?

We tackle these questions by taking a ground-level approach, through case studies based on practical experiences. We offer lessons and examples to farmers associations, non-governmental organizations (NGOs), policy-makers, and academics who are working toward this same goal.

Even proponents and practitioners who are already convinced of agroecology's benefits, such as the authors of this book, face many challenges in effectively supporting and spreading its principles and practices, as well as in creating enabling policies. These are not simple tasks. It is useful to start with a realistic assessment of where we currently stand in trying to accomplish them. Too often, valuable agroecological farming practices exist as oases of success in a desert of conventional farming strategies and policies. In other situations, farmers have successfully developed one or two agroecological techniques on their land, but have not yet encountered the opportunities or support to more fully realize agroecology's benefits by continuing to innovate and add complementary practices. And far too frequently, government policies and programs either neglect family farmers, undercut their agroecological practices, or intentionally work to "migrate" them out of agriculture.

Understanding the starting points of different farmers and communities and their effectiveness in transitioning to, strengthening, or spreading agroecological farming, is context specific. Individuals and organizations create particular pathways in their local settings. To emphasize this diversity of experience, we draw insights from nine grounded cases from different national contexts around the world. These cases highlight the work of farmers and their organizations, but also that of non-governmental organizations (NGOs) and scientists working with them. The common thread is that each case portrays individuals and organizations who are building pathways to scale agroecological farming by recognizing the importance of on-farm innovation, supporting the autonomy of farming communities, and working toward horizontal transmission of knowledge and the creation of more supportive policies.

The strategies described in each case have their strengths as well as their limitations. We believe it is instructive to learn from both. Some of the cases represent the work of 30 years or more, while others have a shorter life. Some are more community oriented, while others are more explicitly focused on wider policy and systems change. While there is no one-size-fits-all model, we do believe that key lessons and principles can be drawn from these and other experiences. These lessons can then inform and strengthen our work in different contexts and environments around the world.

Since the 1950s, there has been a growing corporate concentration of the ownership of agricultural knowledge, technology development, inputs, seeds, and supply chains. This commercial expansion of industrialized agriculture too often displaces the logic of farmer-centered agroecology. People and nature are removed from the center of the equation, and the space for expanding agroecology, and the agency of farmers and their organizations, shrinks. The structures to support agroecological development, as well as the practices of agroecology, need to be strategically combined in order to reverse this destructive trend.

Box I.1 Groundswell international

Groundswell International is a global partnership of NGOs, local civil society organizations, and grassroots groups that strengthen rural communities in Africa, the Americas,
and Asia to build healthy farming and food systems from the ground up. We work with
rural communities to improve their lives by strengthening and spreading agroecological
farming and sustainable local food systems. We support farmer innovation and farmer-to-
farmer spread of effective solutions; strengthen farmers' and women's organizations and
grassroots movements; document lessons; and amplify our voices locally and globally to
shape policies and narratives to nourish people and the planet. Groundswell is currently
working with partners in Burkina Faso, Ghana, Mali, Senegal, Ecuador, Haiti, Honduras,
Guatemala, Nepal, and the United States.

Groundswell International collaborates directly with the organizations and programs
featured in this book from Burkina Faso, Ghana, Mali, Haiti, Honduras, and Ecuador.
In addition, we have invited allies from Brazil, the US, and the Netherlands to also include
their valuable experiences.

Box I.2 Scaling agroecology: Why and how

We believe there is an urgent need to make agroecology central to and widespread
within our farming and food systems, or in other words, to "scale" it. Perhaps the most
important reason to strengthen and spread agroecology is because it works, particularly
for smallholder family farmers. It improves the multiple functions of agriculture, including
food production, income generation, employment, cultural maintenance, environmental
services, biodiversity, and resilience. Again, multiple studies have already documented
and convincingly demonstrated the effectiveness of agroecology, as illustrated, for
example, in Box I.3.

Box I.3 Study results: Scientific evidence supporting the agroecological agriculture model

*"Today's scientific evidence demonstrates that agroecological methods outperform the use
of chemical fertilizers in boosting food production where the hungry live—especially in
unfavorable environments ... Recent projects conducted in 20 African countries demonstrated a doubling of crop yields over a period of 3–10 years ... We won't solve hunger
and stop climate change with industrial farming on large plantations. The solution lies in
supporting small-scale farmers' knowledge and experimentation, and in raising incomes of
smallholders so as to contribute to rural development ... If key stakeholders support the
measures identified in the report, we can see a doubling of food production within five to
ten years in some regions where the hungry live."*
Olivier de Schutter, UN Special Rapporteur on the Right to Food (de Schutter, 2010).

In addition to analyzing agroecology as practice, science, and movement,
the cases in this book highlight the value of supporting strategies to amplify
agroecology at three levels: depth, breadth, and verticality. In practice, the
levels are overlapping and inter-related, and strategies at each level are linked;
however, attention to each level is necessary in order for agroecology to be
more widely disseminated and adopted. The following "scaling framework"

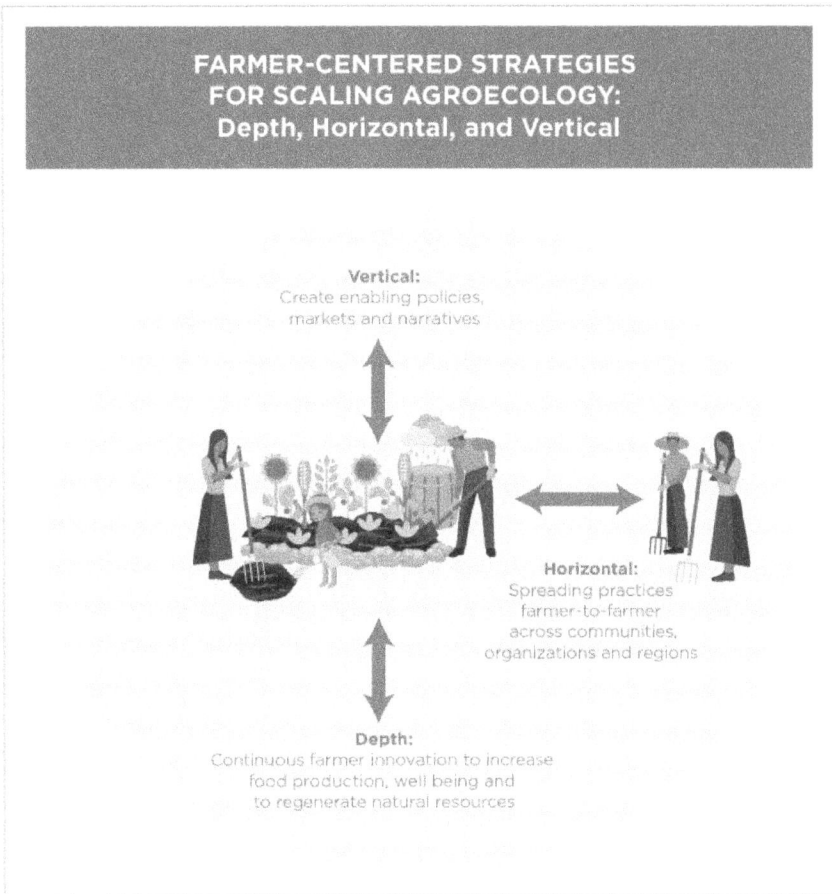

FARMER-CENTERED STRATEGIES FOR SCALING AGROECOLOGY: Depth, Horizontal, and Vertical

Vertical:
Create enabling policies, markets and narratives

Horizontal:
Spreading practices farmer-to-farmer across communities, organizations and regions

Depth:
Continuous farmer innovation to increase food production, well being and to regenerate natural resources

Figure I.4 Strategies for scaling graphic.

can help to capture lessons from each case, and also used to understand other experiences:

- **Depth:** This level is reached when farmers or groups of farmers are able to continuously innovate on their own farms, moving from limited use of agroecological practices to more fully agroecological farming systems that enhance family well-being.
- **Horizontal Scaling (or Breadth):** Agroecology achieves breadth when the principles and practices are horizontally spread across many farming households and communities.
- **Verticality**: This level comes into play when work is done to create an enabling context for agroecology by strengthening wider networks or movements, linking farmers more beneficially to local markets, and creating supportive policies.

Transitioning to healthier farming and food systems

In developing this collection of cases, we had significant debates about the concept of scaling. How does one scale the process of people working creatively and productively with nature, in a way adapted to each local context? Who does the scaling? Development agencies and corporations often use scaling to describe the creation of technologies or practices, and their promotion for adoption or sale to wider populations. This often implies homogenization and uniformity. Industrialized agriculture is well-suited to this type of scaling because it is inherently geared towards standardized, homogeneous inputs and outputs that can be promoted through uniform technology packages. These crowd out biodiversity and local decision-making and control.

When we speak of "scaling agroecology" in this book, what we mean is spreading *an approach to farming and a process for supporting continuous agricultural innovation*, rather than just a specific set of technologies. We understand that farming is multifunctional, and that many agroecological practices are based on long-standing practices and knowledge developed by indigenous people and smallholder farmers, and thus not well-suited to uniform processes of dissemination. The implication is also that we must change not simply the methods of farming, but *the ways that farming is supported by agricultural ministries, philanthropies, NGOs, and scientists*, so that farmers can continue to advance their context-specific processes of innovation. We are talking about a fundamental transition to healthier farming and food systems.

We invite readers to reflect on the experiences and lessons of the organizations and people portrayed in this book as they work to build agroecological alternatives in their own contexts. In Brazil, the Borborema Pole territorial organization is building upon farmer innovation to create a new paradigm for "living with the semi-arid." In Honduras, a 40-year-old agroecology movement struggles to create space for positive change in the face of an undemocratic government. A group of peasant associations in Haiti are building democratic processes and regenerating rural communities. In Ecuador, a collective of organizations are linking rural and urban communities to invest their local food budget in healthy, agroecological production. A farmer and scientist duo in California are creating a model for the transitioning of strawberry production in the US to be organic and socially just. The *Barahogon* association in Mali is recovering their traditional role in regenerating trees to improve their Sahelian landscape. In Burkina Faso, Tani Lankoandé tests strategies to restore organic matter to her hardpan soil, working to make it productive again, and teaching other women to do the same. Farmers, women's associations, traditional chiefs, and civil society organizations in Ghana are linking arms to protect farmers' rights to produce and save seeds. Finally, in the Northern Frisian Woodlands of the Netherlands farmers collaborate with scientists to develop "closed-loop farming."

As societies, we make choices to create the farming and food systems that we have. To overcome poverty, hunger, and climate change, we need to make choices, such as those portrayed here, to shift from business as usual to a new paradigm that strengthens the creative power of family farmers and their organizations and spreads agroecology for healthier farming, food systems, and economies.

Notes

1. For example, The Economics of Ecosystems and Biodiversity (TEEB) for Agriculture and Food is a global initiative focused on "making nature's values visible" and mainstreaming these into decision-making at all levels. See: http://www.teebweb.org
2. Syngenta is one of the world's largest agribusinesses, producing agricultural chemicals (herbicides, fungicides, insecticides, and seed treatments) and seeds (including hybrid and genetically engineered types).
3. The IAASTD report was sponsored by the United Nations, the World Bank, and the Global Environment Facility (GEF). Five UN agencies were involved: the Food and Agriculture Organization (FAO), the UN Development Program (UNDP), the UN Environment Programme (UNEP), the UN Educational, Scientific and Cultural Organization (UNESCO) and the World Health Organization (WHO).

Borborema
Territory
PARAÍBA

NORTHEAST
REGION

B R A Z I L

PARAÍBA

Recife

Salvador

BRASILIA

BOLIVIA

PARAGUAY

São Paulo

ARGENTINA

ATLANTIC OCEAN

Regions referenced in chapter

CHAPTER 1

Peasant innovation and agroecology in Brazil's semi-arid region

Paulo F. Petersen

Summary: *Over many years, peasants in Brazil's semi-arid region have developed ways of "living with the semi-arid," as opposed to "fighting the drought." The case of a cistern—invented by a farmer and disseminated to hundreds of thousands of families in the region—illustrates ongoing processes of farmer innovation. This chapter, by the executive coordinator of Agricultura Familiar e Agroecologia (AS-PTA), describes how civil society organizations have worked to link peasant innovations like this to new forms of local organization in order to create locally generated alternatives to top-down State policies and development programs.*

A farmer innovator

Manoel Apolônio de Carvalho, better known as Nel, is a family farmer from Sergipe state in northeast of Brazil. His life story is similar to those of tens of thousands of other rural inhabitants of the Brazilian semi-arid region. Seeking to escape rural poverty and drought, many have tried their luck by migrating to southern Brazil to earn money before eventually returning home to pursue a living as farmers. Nel found work as a bricklayer in São Paulo in the mid-1990s, and learned to make pre-molded slabs of cement for use in the construction of swimming pools. After returning to Sergipe, where he again faced the challenge of farming under dry conditions, he decided to try the technique for constructing slab cisterns to capture and store rainwater. The result was a cistern that is cheaper and more resistant than the traditional brick cisterns used in the area. The technique quickly attracted interest.[1] Soon, people in and around his community asked Nel to build cisterns for them, giving him opportunities to gradually perfect his invention, while allowing other farmer-bricklayers to learn and train with him. Nel's cisterns are made of cement and iron, constructed by local bricklayers using materials bought in local and regional markets. The Nel-type cistern strengthened the local economy and increased employment. Because installing the cisterns requires excavation, people helped build each other's cisterns, strengthening social capital. The approach was flexible; cisterns could be designed and scaled by farmers according to their local conditions, needs, and resources.

Photo 1.1 The type of cistern innovated by Nel in Sergipe, Brazil.
Photo Credit: Paulo Petersen

Little did Nel know that his adaptation of a technique for swimming pool construction that he learned in the country's biggest and wealthiest city would eventually reach millions of poor people in a semi-arid region of Brazil and help address one of their most vital needs. Nel's innovation was one seed among many that eventually grew into the *One Million Rural Cisterns Program* (1MRC, or *1MCR* in Portuguese). The 1MRC is a regional initiative conceived and executed by the Brazilian Semi-Arid Alliance (ASA), a civil society network comprised of more than 1,000 organizations active in the region's eleven states. The ASA promotes the paradigm of "living with the semi-arid" instead of "fighting the drought." Thanks to funding secured through partnerships with the federal government, private companies, and international agencies, 1MRC built more than 589,000 cisterns for more than 2,500,000 people between 2003 and early 2017. The program has won national and international awards for improving the quality of life in Brazil's semi-arid region.

The 1MRC succeeded because it reproduced the procedures adopted by Nel and his companions through *capacity building* for local bricklayers, so that the knowledge is independently put into practice and adapted by multiple communities; stimulating *peasant reciprocity* for manual activities (such as digging a hole in the soil to hold the cistern); and purchasing construction materials (cement, sand, etc.) from *local markets*. The combined positive effects of these procedures extended far beyond the program's direct impacts on the food security and health of rural families. Above all, the local population was convinced that they themselves could contribute powerfully to development in their region, rather than seeing development as a gift that comes from elsewhere.

A new contradiction

Despite the efficiency of the slab cisterns and the effectiveness of the grassroots methodology of the 1MRC Program, in 2010 the federal government tried to accelerate the impact of cistern-dissemination using a different approach: a massive program offering tens of thousands of new polyethylene water cisterns to the semi-arid region. But unlike Nel's cistern, the new cistern was not made on site. The program eliminated employment for local bricklayers as well as the local purchase of materials, producing a negative multiplier effect on local economies. The State paid for everything, making voluntary labor redundant and eliminating the catalyzing social dynamic of neighbors building cisterns together. Ironically, this also made the new cisterns more expensive. Moreover, the new plastic cistern program lacked flexibility, as farmers couldn't adapt these pre-fabricated cisterns to local conditions.

The new cisterns once again turned "development" into a blessing from above. Among social movements who had helped lead the 1MRC program, this new government program was immediately understood as an expropriation—not only of the Nel-type cistern, but above all, of the grassroots political space that they had constructed. After decades of work by farmers groups, women's groups, NGOs, civil society, and state actors, the ASA and their PIMC program had created positive development alternatives and greater democratic participation for the region. While Nel's cistern had functioned as a seed of social change supportive of ASA's work, the new cisterns acted more like an herbicide.

In response, about 15,000 farmers from across the semi-arid region travelled to the town of Juazeiro, Bahia in December 2011 to mount a large-scale protest against the new cistern program. This demonstration of collective strength led the government to renegotiate its position. Since then, the 1MRC program has coexisted alongside the new government program. This is just one example of the on-going experience of farmers and social movements that support locally generated innovation based on an alternative paradigm.

Top-down development versus grassroots innovation

The story of Nel and his innovation reflects an extraordinary situation that is, at the same time, common to the rural world. It is commonplace because farmers and their organizations generally do not remain passive in the face of oppressive realities. They are actors with social agency, and they exercise it.

However, it is also extraordinary because this farmer agency is typically neglected in public policy and development practice. Nel's innovation, in contrast, was widely recognized and valued through a public program, 1MRC, conceived and executed by a civil society network with an extensive grassroots presence in Brazil's semi-arid region.

The program combined the two complementary aspects of Nel's innovation: the slab cisterns (the hardware) and the social organization to construct the cisterns (the software). In contrast, in its linear and top-down approach, the government sees farmers as individual, passive recipients of public programs. This downplays farmers' creative capacities to combine local resources, both material and non-material, and to solve locally defined problems. This top-down approach exacerbates their dependency on outside solutions, and ignores farmers' innovations while neglecting the potential of their social agency in rural development.

Living with the semi-arid versus fighting the drought

The Brazilian semi-arid region is one of the largest and most populated of its kind on the planet. It covers a geographic area of 980,000 km2, concentrated in 11 states located in Brazil's Northeast. About 22.5 million people live in the region—which is 12 percent of the national population—and 44 percent of these live in rural areas, making it the least urbanized region of the country (IBGE, 2010). The semi-arid region contains more than half of Brazil's population living in poverty (58 percent).

An image of the semi-arid region has developed in the national consciousness, due to the abysmal social indices and recurrent droughts, as a region "historically destined" to be poor and backward. In some conservative intellectual and political circles, it is considered a "problem region." These perspectives induce passivity among the population and also function as a powerful ideological lever for public interventions informed by the notion of *combatting the drought*.

Since the beginning of the twentieth century, the government strategy to combat drought has essentially been to build large hydraulic infrastructures to capture, store, and transport huge volumes of water. As a result, water resources are concentrated in just a few localities, often large farm estates, and fail to meet the water needs of geographically diffuse rural communities. The concentration of water and land ownership reinforces the unequal social structures of power in the region, making poorer populations more vulnerable to the unpredictability of the climate.

Contradicting the fatalistic perspective of *combatting the drought* with outside solutions, generations of peasant families and rural communities from the semi-arid region have developed viable, sophisticated organizational and management strategies for their agroecosystems. They exercise their creativity by innovating and improving their existing farming systems, based on living intimately with the "unwritten codes of nature" rather than fighting against them. They have created mosaics of agrobiodiversity analogous to natural ecosystems, and reproduced environmental services needed to maintain fertility. Their practices have helped to create the alternative paradigm now called "living with the semi-arid" (Silva, 2006; Galindo, 2013; Pontel, 2013).

For a long time, the knowledge, technologies, and social processes created by this movement for peasant innovation went unnoticed or under-valued by public development programs. It was only from the 1980s onwards, with the return to democracy in Brazil, that civil society institutions became structured to provide systematic advice to peasant organizations, looking to associate the critique of the historical pattern of agrarian occupation in Brazil and the conservative modernization project with the building of alternative styles of rural development. Today identified with the agroecological field, these civil society organizations work in integrated fashion with the decentralized dynamics of rural development in the region.

A powerful example was the process developed within the Borborema Territory, located in the Agreste region of Paraíba state.

A history of peasant innovation in Borborema Territory

Within the wider semi-arid region of Brazil, the Borborema Territory— considered the breadbasket of Paraíba—is characterized by a dense mosaic of family farming. Situated between the coast, dominated by vast sugar cane plantations, and the dry *sertão* rangelands, the region's history is marked by periods of "depeasantization" and "repeasantization." These cyclical processes are caused by the changing demand for labor of agrarian elites, who exploit portions of the territory in response to rising or declining demand for agricultural products (Silveira et al, 2010).

From the early 1900s, peasants and large estate owners engaged in endless disputes over the possession of agricultural land. In addition, the properties of smallholder farmers were constantly fragmented as they sub-divided parcels when passing down land to the next generations. As a result, over decades, family farmers had less and less land available to ensure their livelihoods. To survive they had to transform the fertility of their agroecosystems. Farmers gradually reduced and eventually abandoned the practices of slash and burn and fallowing, and adopted management strategies focused on agricultural intensification.

In a long-term study of the transformations in the technical management of agroecosystems in the Agreste region of Paraíba over a 70-year period, Sabourin identified and described the endogenous process of innovation rooted in farmers' socio-technical networks, grounded in relations of shared knowledge, proximity, and reciprocity (Sabourin, 2002). In this way, farmers produced and redistributed both products and knowledge. In his studies of Brazilian rural society, the author also observed that the more peasant farmer communities are closed off, dominated, and marginalized, the more isolated, discrete, or invisible their innovations become (Sabourin, 2009). These observations highlight the importance of territorially embedded collective action in order to create dense social networks of farmer innovation.

In the 1930s and '40s the Brazilian government passed labor laws allowing trade unions. In the 1960s, these laws were extended to create rural workers unions, which initially had a corporatist structure with heavy state management. In the early 1990s, as the rural trade union movement declined, family farmers in the Agreste region of Paraíba formed social movements of resistance and struggle. In response, the Solânea, Remígio and Lagoa Seca rural trade unions took on the challenge of addressing the main problems faced by family farmers in the region. They wanted to connect their traditional political agenda, until then heavily influenced by the national union movement, with the realities and interests of family farmers in the territory.

The emergence of the *Polo da Borborema* and the role of AS-PTA

The result was the development of the Borborema Union and Family Farmer Organization Pole (also referred to simply as the *Polo da Borborema*). The Pole emerged as a collective actor at the regional level in Agreste to help revive and strengthen the pre-existing farmer networks for both social and technical agricultural innovation. This new dynamic was encouraged by a new partnership with AS-PTA that, using an agroecological rural development approach, began providing regional support and advice to family farm organizations in 1993. In order to kick-start the work, AS-PTA supported the unions with participatory rural appraisals to jointly produce knowledge with farmers on the reality of family farming in the region. Farmers also experimented with technical, organizational, and political innovations.

In its studies of family farming agroecosystems in the region, AS-PTA identified three core principles for these innovation processes. First, the maintenance of highly functional biodiversity in agroecosystems, where exotic and native plant species are combined in time and space and perform different functions. Production practices are deliberately designed to optimize the ecological and economic efficiency of the system. Second, storage and management of stocks such as water, seeds, animal fodder, food, capital, etc. This allows peasant farmers in the Paraíban Agreste to cope with the irregular supply of rainwater for agriculture in the region. And finally, the productive intensification of limited spaces. The farmers create areas of high biological productivity, such as household yards and land along the shores of reservoirs. Despite their small size, the productive intensification in these areas play a decisive role in offering food for self-consumption, for sale, or for animal fodder.

Scaling-out through knowledge sharing and experimentation networks

Encouraged by participatory agroecosystem appraisals and farmer-to-farmer visits held inside and outside the territory, around five thousand farm families engaged in innovation processes on their own land and with their own

Photo 1.2 Farmers experimenting with agroecological potato production.
Photo Credit: Paulo Petersen

communities. These joint exercises in knowledge production focused both on agricultural production strategies (such as the diversity of cultivated beans, livestock breeding systems, water resource management strategies, the use of native fruits and medicinal plants, the productive management of household yards, and the use of bio-fertilizers), and methodological and political strategies (including the participation of poor families in the innovation networks and understanding the impact of public policies on the sustainability of regional family farming as a whole).

Exchanges and experimentation networks became an important way for farmers to build technical, organizational, and political capacity. They became *farmer-experimenters:* someone with a problem who has an idea about its cause, and decides to test a way of solving it using locally available resources. They were part of an emerging movement for agroecological innovation within their community organizations and farmer unions.

In a few years, a large range of innovative practices were developed or adapted and incorporated into local agroecosystems. Table 1.1 presents the combinations of traditional farming practices and innovative techniques developed or improved through networks of agroecological experimentation.

Political and territorial development organization

Successful experiences in water resource management and community seed banks gradually spread, stirring the interest of other unions and family farming organizations in other municipalities of the Paraíban Agreste. People saw that

Table 1.1 Relations between agroecosystem management principles and traditional and innovative practices

Management principles	Practices	
	Traditional	*Innovative*
Maintenance of high functional biodiversity	• Consortia and polycultures • Use of fodder or native species • Use of local varieties • Hedge planting	• Recovery, improvement and propagation of local varieties • Evaluation and introduction of new varieties and races • Reforestation of farms • Cultivation in rows • Agroforestry systems • Green manure • Vegetable contour lines
Constitution and management of stocks	• Capital investment in livestock • Clay pits, cisterns, stone tanks, etc. • Domestic storage of seeds • Storage of crop residues as a source of fodder	• Community seed banks • Underground dams • Stone tanks • Slab cisterns and paved cisterns • Silage and haymaking practices
Valorization of limited spaces with high biological production potential	• House yards • Intensive planting in low-lying wetlands	• Improved house yards • Underground dams • Stone barriers

Source: Petersen et al (2002)

family farmers using these practices were better able to resist the 1998–1999 drought. This motivated the region's unions within the *Polo da Borborema* to share their innovative experiences in three new municipalities.

This was the first time that the *Polo da Borborema* presented itself as not only a political actor representing its members in dealings with the State, but also as an organizational space unifying family farmer organizations for rural development in the territory. The Polo formed an action strategy centered on two pillars: 1) stimulating local innovation through networks of farmers-experimenters; and 2) elaborating public policy proposals adapted to the socio-ecological characteristics of the territory.

The Pole as a niche of peasant innovation

The legitimization and intensification of farmer innovation associated with the "farmer-experimenters" was key to increasing cohesion between the Pole's member organizations. By coordinating and providing a strategic

---

Box 1.1 Seeds or grains? Farmer experimentation on local maize varieties

Seen by conventional agronomy as less productive compared to so-called improved varieties—which are only accessible via the markets or public programs—local varieties, known as *sementes de paixão* (passion seeds), have not even been officially recognized as seeds, but as grains. To demonstrate the opposite, a team of researchers from the Brazilian Agricultural Research Corporation (EMBRAPA) was invited to support the network of farmer-experimenters to conduct trials over three years comparing the varieties distributed by public programs and the passion seeds. The results unequivocally demonstrate the agronomic superiority of the local varieties in relation to producing grain as well as fodder. Empowered by the research results, farmer-experimenters told public officials they would no longer accept government bodies reducing passion seeds to the status of grains. The practical implication is that the supply of seeds used by family farmers should be ensured through the action of territorialized networks dedicated to using, managing, and conserving local varieties, emphasizing the active role of farmers as stewards of agrobiodiversity (Petersen et al, 2013).

---

direction to the networks of farmer-experimenters in the territory, the Pole helped them achieve relative autonomy from the State's and the private sector's institutionalized knowledge systems. In this sense, the Pole functions as a strategic niche of peasant innovation. Further validation came again in 2012–2013, when increased productivity and resilience of family farming— due to multiple innovations (including the Nel-type cisterns)—allowed families in the Agreste to be far more resilient in the face of the harsh drought of the last half century.

However, the farmers in the Pole do not seek to distance themselves completely from institutionalized science. With the advice of AS-PTA, the Pole coordinated the farmer experimentation process, but also engaged increasingly with academic institutions. Members designed research projects based on the interests of networks of farmer-experimenters—on water, local seeds, livestock breeding, household yard production, market access, and so on. An example related to local seed varieties is summarized in Box 1.1. Farmer-experimenters value the input of academically generated knowledge, as well as the methodological resources of objective science to advance local innovation. These partnerships also legitimize farmer innovation in the eyes of the State.

The Pole as a political actor

One of the *Polo da Borborema's* institutional innovations was the creation of a *territorial focus*. This set it apart from the union movement's past political agendas, which were frequently disconnected from the real demands, potentials, and perspectives of its members. The Pole developed this territorial focus in part by creating connections between the issue-based networks of farmer innovators dispersed *horizontally* across the territory, and the *vertical* relations established with different bodies of the State, through political pressure to influence rural development policies and programs.

This political and institutional innovation proved extremely important in mobilizing public resources in support of local development. It is significant because, traditionally, union movements tend to be fairly insensitive to the social experimentation and the strategies that rise from within them. The leaders of these movements tend to be professionalized in their posts and gradually become disconnected from the grassroots. The Pole however, builds collective knowledge production at community, municipality, and the territory scales. The networks of farmer-experimenters continue to generate practical learning that constantly renews the Pole's political proposals. Farmer-experimenters also work as activists in promoting favorable public policies.

This link between farmer innovators and policy activists is also seen in the struggle of the social movements to defend their One Million Rural Cisterns Campaign. Other examples are the Pole's criticism of State policies for distributing improved and transgenic seeds in the semi-arid region, the creation of programs and campaigns in defense of local seeds, and the support for farmers in their roles as stewards of agrobiodiversity. In addition, the Pole declared its opposition to the state government's initiative of compulsory spraying of insecticides to combat new pests attacking the region's citrus plantations, and proposed an alternative of conducting experiments with natural, non-toxic products (Petersen et al, 2013).

Self-governance and management of local resources

The Pole has also worked with networks of farmer-experimenters to promote the management and sustainable use of local resources indispensable for agroecological intensification (e.g., equipment, labor, knowledge, money, organizational capacity, local seed varieties, etc.).

A key element to identifying, mobilizing, managing, improving, and protecting common goods is the strengthening of social practices founded on reciprocity and mutual trust. Doing so strengthens and sustains regional economic activities by drastically reducing transaction costs, while improving product quality and growing their scale. For example, we see this in the mobilization of knowledge, labor, savings, and credit used in the construction of the Nel-type slab cisterns. Community associations or informal groups created and assumed shared responsibility for managing Solidarity Revolving Funds. New families benefit as families pay back the loans they took out to build cisterns. As of 2003, over 1,380 cisterns had been built and funded via a revolving fund system; 656 of these were "additional" cisterns built using financial resources repaid by initial participating families, which would not have been built otherwise. This means that the Solidarity Revolving Funds mechanism resulted in a 90 percent increase in the number of families benefitting from the funds originally allocated to the territory by the program. Taking into account that cooperative work by community members reduced the unit costs for cisterns by an average of 30 percent, the initial funds invested were multiplied by a total of 172 percent. Had the 1MRC Program been

---

Box 1.2 New institutional arrangements for collective management of local resources

- **Equipment:** Unions and farmer associations belonging to the Pole have organized the collective management of 15 mobile silage machines. Members establish the rules for sharing the machinery, allowing them to process large volumes of fodder from various plant species grown on family farms. This stimulates the planting of fodder species. Around 150 families benefit, with an average annual output of 20 tons of fodder per family
- **Biodiversity:** Farmers organized a network of 65 community seed banks to conserve agrobiodiversity and reproduce seeds, making them available for planting as soon as the rains start. These high-quality local varieties are adapted to local environmental conditions and crop systems, and strengthen families' autonomy and security in crop production. Farmers also organized a network of nurseries to produce tree seedlings (forest and fruit species).
- **Labor:** Processes to mobilize community labor are very widespread in peasant farming regions. As noted, this was used in constructing cisterns to capture and use rainwater, improving decentralized access of families to water and many other tasks related to agroecosystems management.
- **Savings and Loans:** Farmers have developed 150 Solidarity Revolving Funds to purchase equipment and inputs needed to intensify the productivity of agroecosystems: water supply infrastructure, ecological ovens, screens for use in yards, manure, zinc silos, small livestock, etc.
- **Markets:** A network of 13 agroecological fairs in the region's municipalities, as well as collective sales in institutional markets, especially via the Food Purchase Program (PAA) and the National School Meals Program (PNAE), enables family farmers to sell their diverse produce and improve financial returns.

---

implemented by a private company, the resources invested would have been sufficient for the construction of only 506 cisterns at most, in comparison to the 1,380 built at that time (Petersen and Rocha, 2003).

Mobilization, reconnection, and improvement of "hidden" resources

Zé Pequeno, a family farmer from the Agreste region, has said, "The role of our unions is to discover the treasures hidden in our municipalities." This captures the essence of the experience in Brazil's semi-arid region when farmer-innovators and activists become protagonists of rural development at a broad scale.

The key to the success of this regional movement lies in promoting processes of peasant innovation to utilize previously immobilized local resources to generate social wealth and autonomy. Peasant innovation is driven and encouraged by networks of farmer-experimenters, altering pre-existing work routines, building connections, and responding to the problems faced by rural families and communities. This creates horizontal connections between the farmer-experimenters at larger geographic and organizational scales. These scales run from the agroecosystem—where family farms are the locus of farmer innovation—to territorial scales in which peasant innovation networks lead to new institutional arrangements for building and protecting community resources. The shift of paradigms, from the government notion of "combatting the drought," to the farmers' notion of "living with the semi-arid," shows how local actors, including the ASA, the *Polo da Borborema*, and AS-PTA, reframed their reality, increased their political capital, and helped build a new development pathway based on intensification through farmer labor and agroecological principles.

The story of Nel's cisterns described at the outset of this chapter provides an emblematic example of the challenges faced by farmers and civil society organizations in reconnecting culture, nature, and local agency for rural development. While the state's programs stifled local agency, innovation, and social change, the experiences of 1MRC and the Pole show that centering farmer experimentation, innovation, and exchange in a locally grounded process of social organization and development is a more efficient, equitable, and sustainable path to rural development.

Note

1.   As well as reducing the unit cost of a 16,000-liter cistern from US$ 690 to US$ 240, Nel's invention of a cylindrical cistern eliminated weak points at corners of walls of rectangular brick cisterns, where cracks and leaks frequently occur (Petersen and Rocha, 2003:16–18).

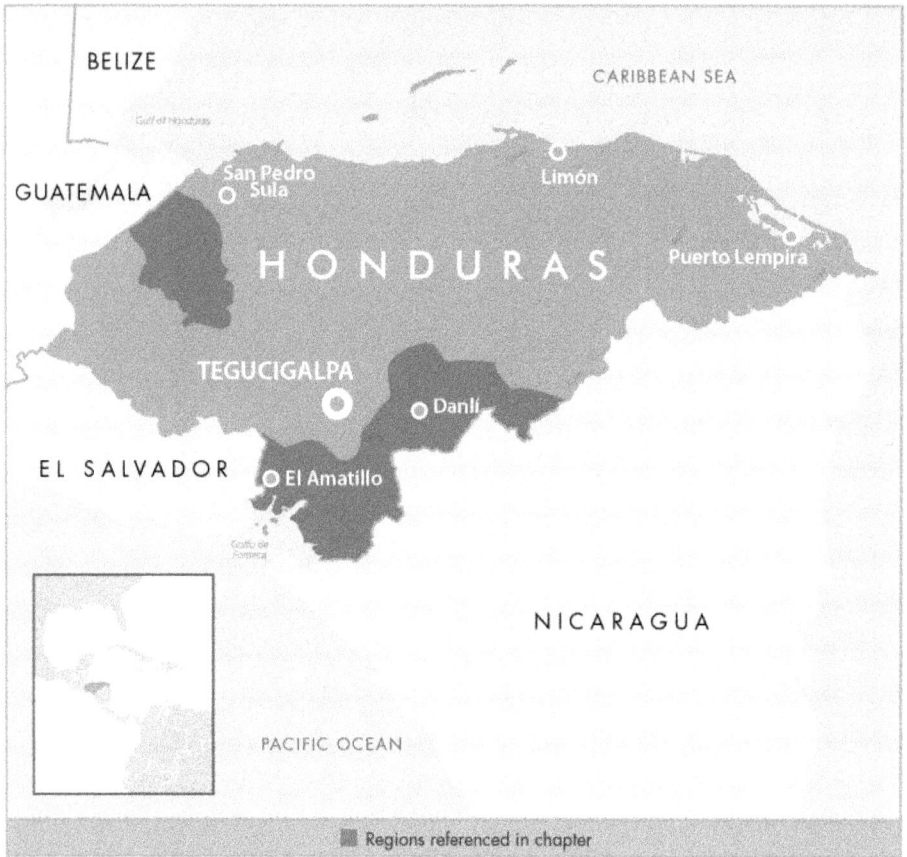

BELIZE

CARIBBEAN SEA

Gulf of Honduras

GUATEMALA

San Pedro
Sula

Limón

HONDURAS

Puerto Lempira

TEGUCIGALPA

Danlí

EL SALVADOR

El Amatillo

Golfo de
Fonseca

NICARAGUA

PACIFIC OCEAN

Regions referenced in chapter

CHAPTER 2

Honduras: Building a national agroecology movement against the odds

Edwin Escoto and Steve Brescia

Summary: *This case study describes the context for the development of Honduras' decades-long movement to create ecologically appropriate farming systems, spread them across the landscapes, and defend the rights of small-scale family farmers. It highlights the work of Vecinos Honduras, an NGO that supports community-led development and agroecology primarily in Honduras' drought-prone south, and the National Association for the Promotion of Ecological Agriculture (ANAFAE), an important network to which it belongs.*

The seeds of a movement

Don Elias Sanchez of Honduras once said, "If the mind of a *campesino* is a desert, his farm will look like a desert." One of the early leaders of Honduras' 40-year agroecology movement, Don Elias sought to improve agriculture by starting with people, not farms. He believed that if farmers' innate creativity and motivation were cultivated (what he called "the human farm"), they could transform their farms, their lives, and their communities.

Don Elias started his career as an educator, and in 1974 joined the Ministry of Natural Resources to direct the training of agricultural extension agents. He became frustrated because while most Hondurans struggled in poverty on steep, mountainside farms, their harsh reality was largely ignored by agricultural professionals. Instead, extension agents promoted the inappropriate technology packages of conventional agriculture (hybrid seeds, fertilizers, and pesticides). He tried to introduce them to alternative thinking and expose them to the realities of rural life through field visits. "'Technology transfer' is an offensive concept," he believed. "You have to transform people" (Smith, 1994).

In 1980, Don Elias left the Ministry to try a different approach, developing his *Granja Loma Linda* teaching farm on the outskirts of the capital city of Tegucigalpa. He turned a tract of poor-quality, steeply sloping land crossed by a ravine into a terraced, diversified, productive farm. It was a place of constant innovation with local resources, where many hundreds of *campesinos* and non-governmental organizations (NGOs) came for hands-on learning about personal development and agroecological farming. Around the same time, he coordinated with the NGO World Neighbors to bring their successful farmer-to-farmer experimentation and extension methodology, developed over the previous decade in Guatemala, to programs in Honduras. From 1980 until his death in 2000, Don Elias is estimated to have helped 30,000 hillside farmers shift from slash and burn farming to more agroecological approaches that were productive and provided a good livelihood (Breslin, 2008).

Then, in October of 1998, Hurricane Mitch devastated much of Central America, causing landslides and massive loss of soil and farmland. It was the region's worst natural disaster in 200 years, affecting 6.4 million people. At *Granja Loma Linda*, a landslide came down the center of the ravine and buried Don Elias' training center and home, even while some of his terraced plots on the hillsides survived. "This Mitch is a lesson I hope we will never forget," said Don Elias at the time (Nelson, 1998). The lesson was that even a model farm could not be protected if the upstream farmers were not practicing soil conservation. Agroecology needed to be scaled up from something being practiced by isolated farmers, to an approach adopted across watersheds and landscapes. Don Elias died in 2000 as the reconstruction of his training center was nearing completion, but his dream lives on.

Photo 2.1 Farmer-to-farmer learning on agroecology continues to this day.
Photo Credit: Alejandra Arce Indacochea

Agriculture in Honduras

Hillside farming, poverty, and political marginalization have long been linked in Central America. As happened throughout Latin America, when Spanish colonizers arrived in Honduras, they took the prime farmland in the valleys, forced indigenous people to work it, and relegated them to hillside farming to produce their own food. It was the beginning of a long and painful history of exploitation and political oppression.

In the 1960s, the "Green Revolution" came to Central America. It was instrumental in creating smallholder systems across the region that combined traditional shifting cultivation (slash and burn) with dependence on modern chemical fertilizers, pesticides, herbicides, and hybrid seeds. As land becomes more scarce and fallow periods shorter, farmers mine the soil nutrients, leading to increased soil erosion, dependence on chemical fertilizers, and a steady decrease in yields.

The conditions of poverty and marginalization in the 1970s led peasant organizations to demand land reform—with some limited successes (Boyer, 2010). Initially, their primary focus was on obtaining access to land, as well as agricultural inputs. Later, a strong focus on agroecological alternatives to conventional farming developed.

In the 1980s, the neoliberal policies promoted in Honduras and across Latin America pushed to "modernize" the agricultural sector. Structural Adjustment Programs (SAPs), required by the IMF, World Bank, and the United States, reduced the role of the state, slashed budgets for agricultural extension, deregulated international trade and investment, and promoted privatization and corporate investment. Support for land reform declined. The favored agricultural model was the promotion of high value (monoculture) crops for export. Overall, family farmers were not considered to be economically viable, with neoliberal theory holding that growth in Gross Domestic Product would trickle down, creating more jobs for them in other sectors. It has not worked out that way.

After Hurricane Mitch, some hoped that the clear evidence of the vulnerabilities created by conventional agriculture, contrasted with the superior resilience of agroecological farming, would lead to a change in national priorities and policies and increased support (Holt-Giménez, 2001a). Instead, neoliberal policies and investments in conventional agriculture were re-doubled. For example, in 2001, the *Plan Puebla Panamá* was launched as a Mesoamerica-wide initiative to promote infrastructure, such as highways, ports, and telecommunications, in particular for export agriculture and tourism. That was followed in 2005 by the passage of the Central American Free Trade Agreement (CAFTA), which built upon and extended the earlier North American Free Trade Agreement (NAFTA, 1994) between the US, Canada, and Mexico.

A 2009 military coup in Honduras weakened the rule of law and increased political violence and impunity (Frank, 2013). Gang violence, drug trafficking, common crime, as well as political persecution left Honduras with the highest homicide rate in the world in 2012 (United Nations Office on Drug and Crime, 2013). In this context, the Honduran government has continued to deepen this neoliberal policy trajectory: providing broad concessions for international mining and hydroelectric companies that allow rural communities to be dispossessed of their land; promoting laws to privatize seed ownership and introduce GMOs; promoting Zones for Employment and Economic Development (ZEDEs)—which are essentially free trade enclaves within the country, with their own laws and governance.

Predictably, the interests of the economic elite predominate in shaping national policies, while family farmers and their interests are largely ignored. *Campesino* farmers seeking to promote agroecology or protect their land and territory have little political voice. In recent years, dozens of *campesino* leaders involved in land struggles have been killed. Tragically this included indigenous, environmental, and human rights activist Berta Cáceres, the General Coordinator of the National Council of Popular and Indigenous Organizations of Honduras (COPINH, *Consejo Cívico de Organizaciones Populares e Indígenas de Honduras*), who was shot to death on March 3, 2016. The situation perpetuates the grinding poverty and social decline for the people of Honduras. In 2013, 64.5 percent of Hondurans were living in poverty (US$2/day), with 36 percent living in extreme poverty (US$1.25/ day). In rural areas, the extreme poverty level is 50 percent. These poverty figures are mostly unchanged from 2004. Income inequality is extreme, and worse than El Salvador, Guatemala, or Mexico (Gao, 2014). In 2015, chronic malnutrition affected 23 percent of children under five, and was over 48 percent in vulnerable rural areas (WFP Honduras, 2015). At the same time, obesity due to unhealthy eating is a growing problem, with 46 percent of Hondurans over 15 classified as overweight or obese in 2008 (The World Bank, 2011). As people seek to survive, immigration to the US has surged since 2000, with remittances accounting for 15.7 percent of Honduras' US$3 billion GDP in 2012 (Gao, 2014).

Response: A movement for agroecology

Growth and resilience

In this tremendously challenging context, the agroecology movement sparked by Don Elias Sanchez and many other leaders has continued to seek pathways to grow and evolve. Farmers and civil society leaders who have witnessed agroecology's economic, social, cultural, and environmental contributions have been seeking to expand its great potential to contribute to a more hopeful future for the country. Since the late 1970s, many farmers' organizations and NGOs have supported farmer-to-farmer approaches emphasizing the participation and leadership of farmers in all research and extension activities. Key sustainable agricultural techniques have included soil conservation, in-row tillage, crop residue management, cover crops, agroforestry, companion planting, and use of organic fertilizers. At the time of Hurricane Mitch in the late 1990's, an estimated 10,000 farmers and farmer-promoters were using agroecological approaches on their farms across Central America. Yet these represented only a fraction of the more than four million hillside farmers in the region at the time (Holt-Giménez, 2001a; World Neighbors, 2000).

After Hurricane Mitch, a study was carried out to measure the resistance and resilience of sustainable agriculture to natural disasters, in comparison to conventional practices (Holt-Giménez, 2001a). Forty local and international organizations working closely with farming communities in Honduras, Nicaragua, and Guatemala were involved. They formed 96 local research teams that did side-by-side comparisons of 902 agroecological plots against the same number of conventional plots.

Key results of the study were:

1. Agroecologically farmed plots fared better than conventionally farmed plots on key ecological agriculture indicators.
2. Agroecological plots had 28–38% more topsoil (38% in Honduras).
3. Agroecological plots had 3–15% more soil moisture (3% in Honduras).
4. Surface erosion was 2–3 times greater on conventional plots. Agroecological plots suffered 58% less damage in Honduras, 70% less in Nicaragua, and 99% less in Guatemala.
5. Some indicators varied significantly by country. Landslides were 2–3 times more severe on conventional, in comparison to agroecological, farms in both Honduras and Guatemala, but worse on agroecological farms in Nicaragua.
6. Agroecological methods may not have added to resilience when damage originated on unprotected slopes or watersheds upstream. There is a need to work at the level of the wider watershed or entire hillside.
7. Some land, such as steep forested hillsides, may simply be unsuitable for agriculture, and farmers should be provided with access to more appropriate and better land. In Honduras, only about 15% of the land is considered appropriate for farming, with much of the rest more suitable for forestry (World Neighbors, 2000).

Bottom up strategies, government indifference, and opposition

In spite of this clear evidence of the effectiveness of family farmer agroecology, after Hurricane Mitch the Honduran government did not increase its support or alter its unfavorable policies. Nevertheless, Honduran farmers' organizations and NGOs, as well as international NGOs, have continued to support various strategies to spread agroecological farming and create a broader movement. Key strategies have included:

Teaching Farms: The work of Don Elias Sanchez and other NGOs contributed to the proliferation of model farms, which functioned as Centers for Teaching Sustainable Agriculture (CEAS, *Centros de Enseñanza de Agricultura Sostenible*). Successful agroecological *campesino* farmers across the country managed their own farms as teaching and learning centers for others interested in agroecology. A group of 30 farms formed the CEAS Network (*RED-CEAS)* to collaborate in promoting this model and sharing lessons (Breslin, 2008).

Networks and Advocacy: In 1995, a number of organizations came together to form the National Association for the Promotion of Ecological Agriculture

(ANAFAE, *Asociación Nacional de Fomento de la Agricultura Ecológica*). Currently the network is composed of 32 farmers' organizations, NGOs, and secondary schools. Together these organizations work with about 20,000 farming families across the country to strengthen agroecological production.

ANAFAE promotes knowledge-sharing and management for spreading agroecology among its members and other allies through exchanges, conferences, workshops, and joint research initiatives. It also acts as a political space to articulate positions and influence national authorities on issues related to agriculture, the protection of biodiversity and seeds, and food sovereignty. For example, during presidential campaigns candidates often promise initiatives to create a million jobs, which are then never created. ANAFAE disseminated research during recent campaigns demonstrating that strong support to spread agroecology could easily generate a million jobs in the country over four years. In the worst-case scenario, it would guarantee sustainable family food consumption (Espinoza et al, 2013). At the municipal level, ANAFAE and its members have supported the elaboration of public policies to promote agroecology and municipal regulations to protect natural resources. ANAFAE has also analyzed and proposed modifications in the national Mining Law, which is providing mining companies with concessions to the land of many rural communities, to a commission of the National Congress (ANAFAE, n.d.).

Peasant Organizations: As mentioned, Honduras has a long history of *campesino* organizations and coalitions, such as CNTC (National Union of Rural Workers/*Central Nacional de Trabajadores del Campo*) and COCOCH (National Coordinating Council of Peasant Unions/*Consejo Coordinador de Organizaciones Campesinas de Honduras*), that have been involved in struggles for land reform and land rights since the 1970s. Rafael Alegria, one of the leaders of these organizations, was also an early leader of the *Vía Campesina,* an international peasant movement, and became its General Coordinator from 1996-2004. In 1996 *Vía Campesina* framed the concept of "food sovereignty" on a global level, and has increasingly promoted agroecology as one of its core elements. Given the context in Honduras, much of the struggle remains related to access to land and defense of territory against the spread of mines and hydroelectric dams, in the face of widespread intimidation and the murder of dozens of farmer leaders in areas such as the Aguán Valley and among Lenca communities in the departments of Santa Barbara and Intibucá (Kerssen, 2013).[1]

Community-based Programs: A number of NGOs are supporting community-based efforts to strengthen and spread agroecology. *Vecinos Honduras* is one of these, and is also one of the 32 member organizations of ANAFAE.

The role of *Vecinos Honduras* in supporting community-led agroecology

Farmer leaders and professionals who had long been involved with the agroecology movement in Honduras founded *Vecinos Honduras* in 2009.

◯ **Farmer testimony**

Olvin Omar Mendoza Colindre[2] is 35 years old, and lives with his wife Nancy Elizabeth Aguierre Lopez, and their two sons (11 and 5 years old) in Los Claveles #1, Azabache, Danlí, El Paraíso. The area is 1,200 meters above sea level, with a rainy climate and temperatures that range from 60-90 degrees Fahrenheit, and is appropriate for growing coffee and other crops. Olvin, Nancy, and their family participate in the Michael Newman Danlí program of Vecinos Honduras.

"My dream is to have a healthy and prosperous family. Our challenges have always been producing enough food for our family for the year, generating more income, and the low prices the intermediaries pay us for corn, beans, and coffee.

"I feel like I am learning a lot of new things by participating in the activities of the program. My main activity is working on my coffee farm, and to continue to diversify it. I used to use chemicals, but have switched to use natural products, which also means less costs for me in producing. I am doing worm composting now and applying it on my coffee plants. We have diversified a lot—in addition to coffee, we now also have plantains, bananas, avocado trees, achiote, papaya trees, cedar, plum trees, guama, chile peppers, tomatillos, tomatoes, peas, apazote, celery, beans, and corn. I feel happy and blessed by God. I am also applying other organic fertilizers we make. We have also reduced our risk of poisoning from pesticides. Two years ago I became intoxicated with Pirineta, which we used to control weevils in the beans.

"For the last two years, I have been able to produce enough corn and beans to ensure our food throughout the year. We have also been able to earn some additional income by selling agroecologial produce, helping us get out of debts we had incurred for agricultural production. We have been able to improve the physical construction or our house. We still need to come together to find alternatives to selling to the intermediaries who continue to give us low prices.

Photo 2.2 Olvin Mendoza checking insect traps on his agroecological coffee farm.
Photo Credit: Edwin Escoto

"Through this program I have been able to participate in processes of experimentation on my own farm. I have become a farmer leader now. Other farmers seek me

(Continued)

Box Continued

out to share my knowledge about coffee farming with them. I have already shared this knowledge with 14 other farmers in the area. My wife has also been participating in the agricultural activities, as well as learning workshops on improved family relations. Now she is teaching what she has learned [about] gender and supporting youth to a women's group in the community, and is even acting as a promoter to share her knowledge with other communities.

"My coffee plants are healthier. My soil has more organic matter and more ability to hold water. There is a much greater diversity of plants now. Three years ago, in a quarter *mananza* (about .4 acres), I harvested only 12 *quintales* (2,645 lbs.). Last year, after two years of these agroecological practices, I harvested 16 quintales (3,527 lbs, a 33% increase). I hope to do the same this year. Before, in our community many people said these trainings were a waste of time, but now the majority are applying these practices. Now we have created an organization of producers, and we plan to start collectively selling our crops together. We think it is the only way to improve our living conditions."

The organization grew out of the previous work of World Neighbors, when it closed programs in the country.

Vecinos' core program strategy is to strengthen the capacity of community-based organizations to lead processes of local development so people can improve their own lives. While programs emphasize sustainable farming and food sovereignty, they also focus on community health, citizen participation, gender, youth, environmental regeneration, resilience to climate change, and risk management.

As Edwin Escoto has described of *Vecinos*:

> "We start programs through dialogue with communities, usually who are asking for our support. Then we carry out processes of participatory planning with them, so together we can understand their reality, identify priorities, and develop initial action plans. Based on the priority challenges farmers have identified, we then facilitate farmer experimentation with them. Often the initial program staff promoters are experienced and successful agroecological farmers from other communities. Local farmers test a few agroecological practices, such as improving soil conservation, the use of cover crops and green manures, and the diversification of the crops, to improve their farming strategies. The key is for farmers themselves to observe the results, and ideally identify recognizable results and benefits. After a year, as they continue to learn and innovate, community organizations select motivated farmers as promoters to teach others, farmer-to-farmer. They use field days, inviting people to visit a successful farm, learning exchanges between communities, and also provide direct advice and follow-up to other farmers interested in benefitting from more agroecological approaches" (Escoto, 2015).

Community organizations, such as the Association of Experimenting Farmers of San Antonio of Las Guarumas, are strengthened to coordinate work they prioritize related to agriculture, health, and other issues. *Vecinos Honduras* works with community members to promote a critical mass of farmers who are experimenting and adopting agroecological practices. The assumption is that this critical mass of farmers, involved in effective local organizations, can create a multiplier effect to further spread agroecological practices across families and communities. Community-based organizations are linked to wider networks and movements, such as ANAFAE, to address the root causes of poverty and environmental degradation and create more enabling policies.

In addition to sustainably improving farming production, with an emphasis on basic grains and diversification for improved nutrition, other activities that often motivate community members include: savings and credit groups; improving basic household and community infrastructure for sanitation, hygiene and health (latrines, water purification, improved cooking stoves, managing garbage, etc.); local grain banks; and leadership training for women, men, and youth. "In *Vecinos* we are placing more emphasis on working with communities to better connect farmers to local markets," said Escoto. He continued:

> "For example, using community radio and other popular communications tools to spread understanding about agroecological farmer and the value of people eating the local, traditional foods that farmers produce. We believe there is important opportunity for youth, many of who don't see a strong future for themselves in their communities. We are also facing an incredible drought crisis in southern Honduras. To address this we are focusing more attention on water harvesting and water management in the context of agroecology. We are part of a learning network with six other NGOs and farmers groups in the region to share lessons and strategies for confronting drought" (Escoto, 2015).

Results in southern Honduras

In spite of being marginalized, family farmer agriculture remains vital to Honduras' economy and food security. About 50 percent of Honduras' population of 8 million continues to live in rural areas. Few Hondurans realize that family farming households still produce 76 percent of the food consumed in the country—including staple crops of corn and beans. Overall, the agricultural sector employs 37 percent of the working population, and generates 14.3 percent of gross domestic product (GDP) (Espinoza et al, 2013).

Some results from Vecinos Honduras programs

While *Vecinos Honduras'* programs are still young, they allow people to sustainably improve their lives. As of the end of 2015, *Vecinos Honduras* was supporting six programs, working in 65 communities, with over 1,400 families engaged and improving their wellbeing (about 7,500 people). Over 880 farmers (42 percent women) have been specifically involved in farmer experimentation and farmer-to-farmer learning, and are adopting improved agroecological practices. These farmers are regenerating 980 hectares (2,421 acres) of degraded

Photo 2.3 Woman with millet harvest in southern Honduras.
Photo Credit: Christopher Sacco

land. In the context of chronic drought, achieving food security through increased household production is a long-term goal. Initial evaluations show that 20% of the involved families have achieved food security through their own farm production (enough food for the year), while 40% produce enough for 8 months; and another 40% only produce enough for three months.

"We have seen evidence of improved gender relations and empowerment of youth as well," says Escoto. "This includes reduced domestic violence, more sharing between men and women of domestic labor, and increases in female leadership. For example, 58 percent of the membership of community organizations focused on economic activities is women and youth, who occupy 41 percent of leadership positions. At the same time, we see that in the top decision making positions at community levels, women still hold less than 30 percent of those positions. So there is still much work to be done" (Escoto, 2015)

Participatory research by ANAFAE

In 2013, ANAFAE collaborated with *Vecinos Honduras* and other members of the network that are working in Honduras' drought prone southern corridor, to study the impact of agroecological family farming in the area (Escoto, 2015). The study confirmed the great importance of family farming to people in the area, and the superior results of agroecological farming strategies.

A least 80 percent of food consumed weekly by rural or semi-urban families in the area is produced on their farms. Over 50 percent of average monthly incomes of these rural families, or approximately US$275, also comes from their farm production. The study demonstrated that agroecological farmers are more able to meet their food needs, and resist food shortages, in comparison to conventional farmers or low-income urban families. According to the rural families surveyed, agroecological farming also produces more and more varied food reserves than conventional farming. In terms of employment generation at a national level, as noted earlier, ANAFAE determined that if the government were to seriously support agroecological farming, it could generate the equivalent of one million jobs in the country over four years (ibid).

One effective mechanism that rural communities in the south have found to increase access to food in the face of recurring droughts and food shortages is the establishment of community grain banks or reserves. Families with agroecological farms generally enjoy better health because they are less exposed to agro-toxins in their food and have a healthier and more diverse diet, with high consumption of fruits and vegetables. Participants also shared their perspective that agroecology is best practiced as a family project. It requires collaboration, and can contribute to improved gender and family relations if accompanied by complementary reflection and educational strategies. Agroecological farming has increased the interest of youth in their farms. In contrast to conventional farming families, many of these young people see the value in inheriting and managing their family farms.

An interesting finding was that agroecological farming families achieve higher levels of education on average, in comparison conventional farmers in their communities, due to increased motivation and valuation of personal development opportunities.

Finally, ANAFAE's study showed that, in response to the wave of laws passed to promote extractive enterprises, a growing social movement is emerging to protect family and community rights to their land and territories. Agroecological farmers are linking with other social actors to defend their rights to territory.

Key lessons

The movement to spread agroecology and contribute to food sovereignty has been going on in Honduras for 40 years under incredibly challenging circumstances. Yet many farmers are practicing agroecology, and multiple local farmers organizations and NGOs, as well as international NGOs, are supporting it. *Campesino* farmers are still responsible for producing the great majority of the food that Hondurans eat. While documented evidence shows that agroecology has been resilient and beneficial for those practicing it, its full benefits and potential are undervalued by society and neglected or undermined by government. Just as the destructive force of Hurricane Mitch in 1998 exposed both the potential and the limits of agroecology, the wave of neoliberal policies since the 1980s and the deteriorating rule of law since the 2009 coup have made Hondurans more vulnerable. Widespread support for and adoption of

---

💬 **Farmer testimony**

Juan Ángel Gutiérrez3 is 38 years old and his wife Alba Luz is 36. They have six children, ages 18, 16, 14, 8, 2 years, and 2 months old. They live in the community of Caserío del Mal Paso, in San Antonio de las Guarumas, in Nacaome Valle, Honduras. The area is in the dry corridor of southern Honduras, characterized by frequent droughts. His community is 70 meters above sea level, with eight months of dry season and temperatures that range from 82–107 degrees Fahrenheit. Juan, Alba Luz, and their family participate in the Las Guarumas program of Vecinos Honduras.

"I was born here in Caserío del Mal Paso, and grew up here. Since I was a child I have been cultivating corn and sorghum using conventional techniques. I've stopped practicing them because I saw the negative impacts on the environment and on my family. My family has changed since we began participating in these activities, even though I've only been participating for a couple of years. We are involving the whole family in the farming activities, and are using agroecological practices that provide great benefits to my family and the community. Now we have more food than before. But I am also busier than before working the land, because before I only grew corn and sorghum, and the majority of the time I spent playing billiards. Now we are also growing beans, cucumbers, yucca, sweet potatoes, and squash for our family consumption. I am thinking about integrating fruit trees into our farm for the longer term. I've been learning about and using organic compost, and [learning] which crops do best under these dry conditions. We've seen a lot of positive changes. We have a family garden that is producing, an improved wood saving stove, and a latrine.

"We have a great challenge with lack of water here for drinking, household use, and to irrigate crops. We've had a great challenge producing enough food for our family. We have constructed a tank from recycled materials to collect rain water. And we are treating and managing the limited water better. But we have been participating with the community organization in learning opportunities on agroecology and health.

"But also the communication and relations among my family members has improved. I think what makes me most proud is the unity of our family. Now there is more communication and respect. Our children participate in the youth group. We've diversified our production, and are collecting more water and using it better. Before, our children constantly had respiratory infections. Now our health is better and they rarely get sick."

---

agroecology could increase resilience to climate change, drought, and disasters; improve the wellbeing and health of millions of people; generate employment; and contribute to a more democratic and just society.

A number of lessons can be drawn from the experience of the agroecology movement in Honduras. First, farmer experimentation and farmer-to-farmer learning are key to spreading agroecology. The first teaching tools that farmers use are their own plots, their knowledge, and their own words. In addition, agroecological farmers first produce food for household consumption, and then for the sale and exchange of surplus products.

Second, agroecological farms are more resilient to climate change and natural disasters, but agroecology must be adopted across wider landscapes to reduce vulnerability. Agroecological production systems both enhance diversification on farms and depend upon the biodiversity of healthy local seed systems. This genetic diversity is threatened by the promotion of GMOs, which displace biodiversity and create dependence on inappropriate chemical inputs. Agroecological farming also improves family well-being and reduces the motivation to migrate.

Says Edwin Escoto:

"Agroecology is really a way of life. Most successful agroecology experiences are not just about changing agricultural practices, but are rooted in peoples' values. For some farmers, this is a shift in worldviews. There can be a shift from an extractive, short-term perspective, to one that is more long-term and regenerative. A challenge is that it may take several years for farmers who are making this transition to observe the full benefits. For many people, we see a commitment to agroecology also includes a spiritual commitment to personal, family, and community development. It is more than technical. As agroecology is a family project, we see how it also provides opportunities to reflect— and to act—on improving relations between men, women and young people. We also see that many agroecological farmers become local leaders and more engaged citizens. They are working for healthier communities. They are working for a more just and democratic society. All of these things are connected" (Escoto, 2015).

Next steps

Neoliberal policies of the government and international agencies, promoting extractive mining, hydroelectric, and large-scale monocropping projects are threatening family farmers' rights to the land and territories they depend upon to live. Alliances are growing among community organizations, NGOs, and farmers' organizations to defend their land rights and change these policies. Organizations in Honduras need to make the public, the consumers, and the politicians more aware of the realities and benefits of family farming and agroecology. Most Hondurans do not realize that an estimated 76 percent of food consumed in the country comes from peasant agriculture, and that spreading agroecological farming can generate many economic, social, environmental, and cultural benefits. It could also generate significant employment and income generation for family farmers.

Even after the clear evidence of agroecology's effectiveness in the wake of Hurricane Mitch, farmers' organizations and NGOs have not yet adequately succeeded in influencing policies (Holt-Giménez, 2001b). This seriously limits the degree to which agroecology can be spread, and leads to persistent poverty and vulnerability. Supportive policies, as well as strengthened rule of law and democracy, must be created to allow agroecological strategies and movements to fully contribute to a more beneficial future for Honduras.

Notes

1. At least 53 peasant leaders were documented as murdered in the context of the agrarian conflict in Aguán between Sept. 2009-Aug. 2012
2. Mendoza, Colindres, Olvin Omar. Interview, August 14, 2016.
3. Gutierrez, Juan Ángel. Personal Interview, August 5, 2016

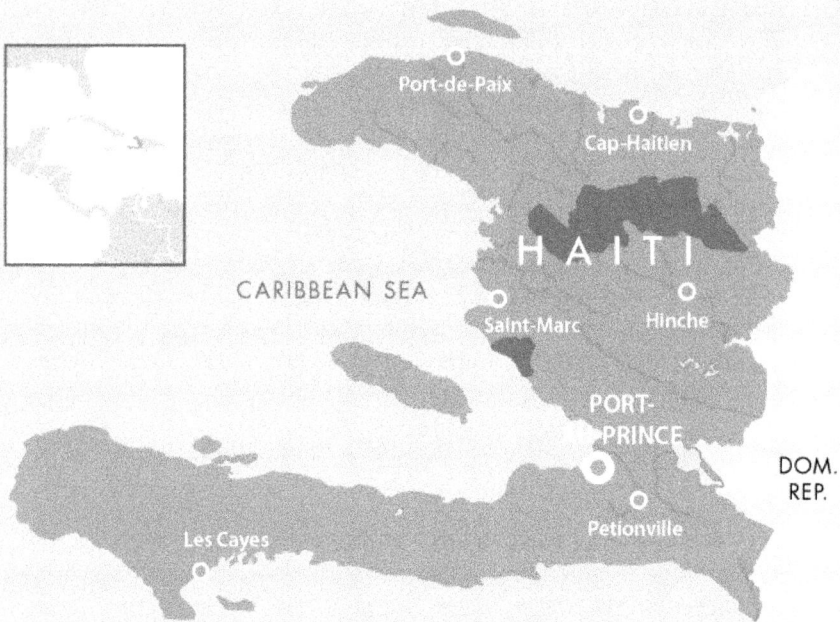

CUBA

CARIBBEAN SEA

Port-de-Paix

Cap-Haïtien

H A I T I

Saint-Marc

Hinche

PORT-
PRINCE

DOM.
REP.

Petionville

Les Cayes

Regions referenced in chapter

CHAPTER 3

A foundation for Haiti's future: Peasant associations and agroecology

Cantave Jean-Baptiste and Steve Brescia

Summary: *Partnership for Local Development (PDL in French) strengthens peasant organizations from the bottom up to create democratic participation and agency in spreading agroecological farming. In a political context of dysfunctional government and extremely weak institutional capacity, this contributes to the creation of decentralized development and the regeneration of degraded land and rural livelihoods. This study emphasizes the importance of local social structures for supporting the spread of agroecological innovation.*

Context: Weak government, deforestation, and rural-urban migration

"A long time ago, everywhere you see a coconut tree had a house next to it." Jean Louis Valere, farmer and community leader, looks across the barren and rocky mountainsides of his community of Bois Neuf, in Haiti's North Department. "Life was really beautiful," he says. "But people left primarily because the land couldn't produce anymore, due to the lack of trees. And now we have soil erosion. People had to go down to the cities—cities of houses on top of houses. But if there were an improvement in the land, then people could come back and build their homes in the villages again."

Jean Louis' statement captures the vulnerability of Haiti's people due to a history of extreme soil erosion and the degradation of natural resources. That vulnerability was dramatically exposed on January 12, 2010 when, in the space of minutes, a 7.3 magnitude earthquake struck the country. The dilapidated "houses on top of houses" in the capital of Port-au-Prince and surrounding cities, over-crowded with migrants from rural communities, collapsed and killed over 220,000 people while injuring another 300,000.

If Haiti is to ever create a more resilient and hopeful future, it must do so on a rebuilt foundation of thriving local communities. The land and rural livelihoods must be regenerated and restored, and the historical flow of people and resources to Port-au-Prince reversed. After the earthquake, *Partenariat pour le Développement Local (PDL,* Partnership for Local Development), a Haitian non-governmental organization (NGO), and its international partner organization Groundswell International, re-committed to a vision and a plan to work with rural communities to build this alternative.

The 2010 earthquake was only the latest and most shocking disaster to expose Haiti's vulnerability and exacerbate its already deep poverty. Following Haiti's independence in 1804, the legacy of slavery and colonial rule, combined with international manipulation and a long string of corrupt and repressive governments, turned what was once the lush "Pearl of the Antilles" into a land that is 98 percent deforested. Since the time of colonialism, governments have functioned primarily to extract resources rather than promote development. Today, government institutions and democratic processes remain incredibly weak, with rural communities receiving almost no effective state support. Floods, hurricanes, and droughts regularly plague the country. Since well before and after the earthquake, far too many development programs, both large and small, have failed to generate lasting results. Many have contributed to greater dependency.

A strategic response: strengthening the agency of peasant associations

"PDL works first to strengthen the capacity and agency of family farmers and peasant associations to manage their own development processes, in a way that is not dependent on external programs," says Cantave Jean-Baptiste, PDL's executive director. "We want peasant associations to be able to say, 'we are not empty bowls waiting to be filled by handouts. We are actors. We are human beings. We have capacity. Here is what we have already done, and here is what we plan to do in the future.' So this involves a process of constantly renewing and broadening the base of local leadership among women, men, and youth.

Photo 3.1 Cantave Jean-Baptiste meeting with *gwoupman* members in the Haitian countryside. *Photo credit*: Steve Brescia

It means strengthening healthy democratic organizational structures from the bottom up. It is these peasant associations, then, who work to spread agroecological farming, strengthen family livelihoods, build local economies, and promote community health."[1]

PDL initiates work with rural communities, beginning with an analysis of existing community organizations. In many cases, these community organizations primarily operate to channel donations from charities—with power concentrated in the hands of one or two "big men." Once they understand the starting point, PDL staff use participatory methods involving a wide representation of community members to facilitate a shared critical analysis of existing community assets and to diagnose priority problems as well as viable, accessible opportunities to improve wellbeing. Based on the communities' own analysis, PDL then fosters an organizational structure for cooperative action from the bottom up. This is done on three organizational levels:

1. *Gwoupman* are solidarity groups of 8–15 women and men, based on collective action, trust, and reciprocity. Each *gwoupman* mobilizes their own resources in a small joint savings and credit fund called *zepoul*, literally meaning "chicken's egg" in Haitian Creole. Rather than "consuming the egg," members work together to invest this initial fund in sustainable farming and economic activities to multiply it in ways that improve their lives. In the process, the local leadership base is broadened as members gain new practical and organizational skills. As described in the farmer testimony box below, membership in the *gwoupman* is not only economically and practically beneficial for community members, it can also instill them with a sense of pride and leadership.

2. Blocks are village-level committees that link together 3–5 *gwoupman* in a community. They coordinate activities among *gwoupman*, such as the promotion of sustainable agriculture, savings and credit funds, and community health initiatives.

3. Central Coordinating Committees (KKS in Creole) link and coordinate activities across 10–25 villages, and are led by regularly elected leaders emerging from the *gwoupman* and community levels.

These organizational building blocks make up **peasant associations**, inter-village organizations that typically have 800 to over 2,000 members each, representing a population of 6,000–10,000 people. They take names like the *Union of Peasant Gwoupman for the Development of San Yago*. Peasant associations hold annual assemblies to plan and assess their activities, report on community-mobilized assets (savings and credit funds, seed banks, etc.), and democratically elect leaders.

By working together in inter-village associations, people are better able to address needs that go beyond the capacity of individual families (e.g. preventing cholera, generating savings and credit, preventing soil erosion, promoting

◯ **Farmer testimony**

Silmène Veillard, Mother, Head of Household, and Farmer, Saint Raphael, Haiti.[2]

"In 2011, we began to collaborate with *Partenariat pour le Developpement Local* (PDL) to develop our local organization. Now we call it the Union of Peasant Gwoupman for the Development of Mathurin, (IGPDM, from the Haitian Creole). In IGPDM, we, as farmers, can organize, share knowledge and ideas, and work together to make positive changes.

"I started by testing some easy and affordable farming techniques. Gradually, my farm began to improve. My garden is now beautiful! It is producing much more and I feel like I know how to work the land better. I am able to feed my children well every day now. After participating in some trainings and applying sustainable agriculture practices on my own land, I became an Agriculture Volunteer. Now, I provide services to the other community members on how to better prepare their farms with sustainable, productive practices.

"We were able to dig a latrine, and we also have a water filter to treat drinking water, so we are not getting sick as frequently. That has been important to prevent cholera in this area.

"In September of 2014, I borrowed 2,500 gourdes (about US$39) from our women's savings and credit group to purchase a new goat. The goat has already produced 10 offspring from four pregnancies, and I have sold several for income. I currently have seven goats, three cows, three pigs, and 10 chickens. I bought the cows with profits from selling my crops. I have been able to save 1,300 gourdes (about US$20), which is more than I have ever been able to save. I am also a member of the agricultural volunteers association, and together we were able to borrow money to buy a plow in order to sell plowing services to others.

"Now I am able to send my three children, two boys and a girl, to school. Many other people in our community are working to send their children to school. I have been a member of IGPDM for over five years. Every time I go to a meeting, I am proud to be a member of the *gwoupman*. I feel like I now have more value in the community … People call me if they have a problem, or if they need to make a decision.

"As an organization, we are working towards our vision and have achieved a lot together. We have bought land, constructed a building for our organization, improved our roads that are difficult to pass in the rainy season, are establishing a savings and credit fund to serve the community, and set up a business cooperative. These achievements have great importance for all of us in the community, even for those who are not yet members of the *gwoupman*. IGPDM is open to collaborating with everyone without discrimination. Anyone can access the credit, seeds, and trainings or participate in other activities if they choose. As citizens, we are even working together to meet with the Mayor of Saint Raphael to advocate to improve the roads in our region which are barely passable in the rainy season."

Photo 3.2 Silmène veillard on her farm.
Photo credit: Cantave Jean-Baptiste

reforestation, controlling free-grazing of animals, negotiating productive relationships with other actors, etc.). As the capacities and interests of the peasant associations grow, PDL adapts its support roles. Generally, peasant associations are able to function with a high level of autonomous capacity within five to seven years.

The growth of peasant associations

PDL began working with peasant organizations in 2009, but the fruits of their labor really took off in 2012. From October of 2012 to June 2014, the number of peasant associations grew from 12 to 17, with an expanding overall membership from 14,600 to 24,580 members—a 68 percent increase.

PDL works within a common territory (see map). This facilitates communication and reduces the costs and time associated with exchange visits, learning, and coordination. The 17 peasant associations each represent a different Communal Section (Haiti's smallest administrative unit), and are present within four of Haiti's nine Departments. This territory contains diverse agroecological zones (from hillsides to flatlands), and different land tenure dynamics and relationships with local markets.

Strategies to strengthen and scale agroecological farming

Haiti is characterized by a virtual absence of government extension services. Instead, large-scale government and international donor programs intermittently extend projects and services around the countryside with limited coordination. Their approaches often undermine, rather than strengthen, family farmer livelihoods and agroecology. Peasant farmers have had to self-organize to manage their own agricultural innovation and extension programs.

Table 3.1 Organizational development of farmer associations

Growth in Members	Total (April 2012)	Total (June 2013)	Total (June 2014)
No. of members	14,600	19,901	24,580
No. of female members	Gender breakdown unavailable	10,866 (55%)	13,994 (57%)
No. of male members	Gender breakdown unavailable	9,035 (45%)	11,091 (43%)
No. of Gwoupman formed	956	1,296	1,548
No. of Block Committees formed (village coordination)	128	169	203
No. of Peasant Associations (inter-village coordination)	12	14	17

"The Peasant Associations establish committees within and across villages to coordinate the testing and spreading of sustainable farming practices," says PDL director Cantave Jean-Baptiste.

> "This is how they are able to regenerate farms and improve and diversify production. By coordinating with these community-based organizational structures, it is possible to organize practical training and information sharing sessions across many families in a village, or across 10-20 villages at a time. For example, the farmers come together on one persons' farm to learn how to mark contour lines with the simple "A-frame" apparatus to build soil conservation barriers. Or they learn to select corn seed to improve the quality of local seed varieties. Then they return to their own farms and communities to test these same ideas and see how they work. They adapt them to their local conditions. Some take responsibility as volunteer agricultural promoters to share successful techniques with other farmers. This is how the organizational capacity of the local peasant associations is directly linked to the spread of agroecology. For agroecology to spread, the peasant associations must do the work. This is the dynamic in one peasant association, across 10-20 villages. You can multiply this across the 17 peasant associations we are working with."[3]

While most farmers in Haiti are not familiar with the term "agroecology," PDL has worked with them to develop its principles and a "basket" of effective practices appropriate to the local context. The overarching agroecological principle is to create a long-term balance between smallholder production systems, soil fertility, conservation, and natural resources. These farming strategies build on existing knowledge and practices (e.g. qualities of local crop varieties, diversification, seed saving), while also fostering important changes to existing farming practices (e.g. stopping the traditional practice of "slash and burn" and introducing soil conservation). As alternatives, farmers test and promote a combination of agroecological techniques that address five major issues: control of soil erosion; increasing soil organic matter and fertility; improving access to and management of quality seed; improved on-farm crop diversity and management (inter-cropping, rotation, optimal plant spacing); and improved plot maintenance (e.g., through timely weeding, control of local pests and diseases, etc.). The practices have allowed peasants to develop farming systems that are more productive and are also more resilient to shocks such as droughts, heavy rains, price spikes, and unpredictable rainfall due to climate change. Some feel that Haiti now has two seasons: drought and hurricane.

The Peasant Associations recognize the "model farmers" who adopt a wide set of agroecological principles and practices, and "agricultural volunteers" who provide farmer-to-farmer advice and support to others. Farmers in each association define their own criteria for what it means to be a model farmer. For example, the farmers' association of the *Sans Souci* village has decided that a model farmer has to "make the earth speak." In the village of *Baille*, a model farmer must practice soil conservation; place five anti-erosive structures on each 0.25 *carreau* of land;[4] cultivate a diverse variety of foods, such as sweet cassava, cassava, pigeon peas, sweet potato, yam, ginger, sugar cane, maize, beans, bananas, tarot, eddoes, etc.; produce enough or generate adequate income to be food secure; and plant fruit and forest trees on their farm for food, fodder, fuelwood, and construction.

Often the rapid, recognizable successes by the first group of innovating farmers quickly become evident to others in the community, who are motivated to adopt the same techniques. The agricultural volunteers support other interested farmers, including those who are not members of the peasant associations, to adopt the most beneficial techniques.

It may take farmers one to three years to see the full and sustainable benefits of transitioning to agroecological farming. To support this transition and further incentivize farmers, PDL works with the peasant associations to develop and manage complementary activities.

These include: savings and credit groups (to access credit for basic tools, labor, and other local inputs); seed banks (to access quality seeds); grain storage banks (to store grain after harvest to improve food access in subsequent months, or receive a better price when selling); group income-generating activities (including women's local commerce and marketing of agricultural products); *ti boutiks,* or community managed stores for basic goods; and community health initiatives. The traditional collective work practice of *konbits* has been rejuvenated to mobilize the labor needed for soil conservation and other activities.

***Chouk* Farms:** Farmers call particularly diverse farming systems *chouk* (rooted) farms. Instead of just planting one crop such as maize, they plant a variety of crops, including root crops, tubers, and a variety of trees. Farmers design the *chouk* systems to increase their food security throughout the year and become more resilient to drought. For example, beans can be harvested after two months and stored. Sweet potatoes can be harvested at 2.5 months and for up to five or six months. Cassava can be harvested between 12–24 months after planting and is particularly drought-tolerant. Bananas produce throughout the year. Papaya trees produce fruit after one year, while mangoes require five to six years. Improved *chouk* farms are based on traditional practices that have been displaced by "modern" mono-cropping systems.

Photo 3.3 Farmers in Haiti develop highly diversified farm plots to harvest food crops throughout the year, such as this farmer on his *chouk* farm.
Photo credit: Ben Depp

> **Konbit**: "*Together, a group could hoe all of this land in one morning. Working alone, it could take one person a month ... Before PDL, we each worked alone ... Training is the most important tool to allow people to come work together. There is a big difference. People are able to do more work, and they're producing more too ... All of these soil conservation walls were built by group members to protect the soil ... Throughout Bois Neuf, it seems that everyone wants to become a group member. Once everyone wants to participate, in three years almost all Bois Neuf can be using soil conservation.*"
> —Jean Louis Valere, Bois Neuf, 2014

Results of strategies to scale agroecology

In 2014, a study carried out in the region revealed extensive impacts from the work of PDL and the farmer associations (CFM,2014). From 2009 to 2014, more than 20,545 farmers have learned improved agroecological farming practices. Table 3.2 summarizes some results of the spread of agroecological practices.

In order to diversify livelihood strategies, and in response to limited access to land, most Haitian farmers manage multiple plots. On average, *model farmers* and *agricultural volunteers* applied improved agroecological techniques on two out of three of their plots, while a majority of other *gwoupman* members have applied agroecological techniques in one of three of their plots. In many communities, farmers who are not members of the peasant associations have also adopted agroecological techniques after observing their effectiveness.

An estimated 20 to 30 percent of all the plots within the 17 Communal Sections reached have adopted some combination of agroecological

Table 3.2 Selected results of farmer-to-farmer spread of agroecology

	Cumulative Totals		
Agroecological Extension	*April 2012*	*April 2013*	*April 2014*
Farmers learning and practicing agroecology	7,039	10,409	20,545
Active Model Farmers (MF)	0	5,617	11,510
Agricultural Volunteers Trained (AV)	116	362	646
Seeds Distributed (metric tons) through community-managed seed banks	0	75	150
Farmers Receiving Seeds from community-managed seed banks	2,388	6,466	7,243
Tree Seedlings Planted on Farms from community-managed tree nurseries	213,790	328,702	467,874
Plots Protected by Agroecological Practices *	0	4,119	6,875
Number of Hectares (acres) Improved by Agroecological Practices			678 hectares (1,676 acres)

* Farm plots average 0.10 hectares in size (approximately 0.25 acres)
Source: Conseils, Formation, Monitoring en développement (CFM). *Evaluation of PDL strategies for scaling agro-ecological farming alternatives,* November 12, 2014

techniques. PDL's goal is to achieve a critical mass of 40 percent adoption. This critical mass will gradually spread to more farmers and become the norm through informal learning mechanisms, without the need for structured extension. After five years of work, PDL and the 17 peasant associations are about halfway to achieving this goal.

Why farmers value agroecology

Farmers value agroecological approaches for a number of reasons: they help to increase both crop and forage production; conserve and retain water (through mulch and increased soil organic matter); and are more resilient and productive during droughts. From 2013 to 2014, farmers reported that in spite of drought conditions crop yields increased, most notably for beans: 17 percent in *Mathurin (Saint Raphael)*, by 22 percent in *Sans Souci (Mombin Crochu)*, and 70 percent in *Ivoire (Arcahaie)*. Corn yields increased as well because of soil conservation, soil improvement, better management of plant density, and improved seed quality.

Evaluators have observed the restoration of degraded plots in all communities. The practice of slash-and-burn has decreased greatly, and most farmers report that they now have greater appreciation for the importance of soil organic matter. Farmers with agroecological plots, including *model farmers,* reported increased food production and food security throughout the year, with some producing a surplus to sell in local markets. Increased income from complementary economic activities, supported by savings and credit funds, has also improved access to food for many households.

Activities to raise nutritional awareness have fostered a change in household practices. Families are eating healthier diets of more nutritionally diverse foods grown locally in farmers' plots, particularly leafy green vegetables. Such foods are more accessible and affordable.

The 2014 assessment also revealed significant social benefits that are less tangible. Personal trust and solidarity were strengthened within families and between members, with a reduction in certain traditional taboos or myths that generate distrust. According to a member in *Sans Souci*, "One can rest easy, and go about his or her activities regardless of the hour, day or night. I don't have to worry about property being stolen, or my personal safety" (CFM, 2014). There has been a strong resurgence of traditional work groups *(konbit)*. Members report that they have an improved sense of self-worth and self-confidence in presenting themselves in public, and an improved ability to negotiate constructively with the government, neighboring communities, urban social groups, and private business people.

Creating an enabling context: Vertical scaling

Haiti is a profoundly challenging context for effective rural development. The institutional capacity and legitimacy of the Haitian government remains extremely weak. Because of political conflicts and problems with election

Roland Moncette, Saint Raphael[5]

"My name is Roland Moncette and I live in the countryside of Haiti, in the San Raphael communal section of San Yago. I used to go looking for work in the Dominican Republic, but I don't anymore. My farm is doing much better, and I feel there has been a change in my life because of the work of our peasant association.

"Before, when I kept goats I was obliged to sell them for a low price before school started. Now, thanks to the loan provided to me by my community organization, the Union of Peasant *Gwoupman* for the Development of San Yago, (or IGPDS, from the Haitian Creole), I am able to borrow money for school and leave the goats to keep reproducing so I can earn more. With the increased yields in my garden, I was also able to buy some land and goats. I have started building a small house to live in, and I have more than 10,000 gourdes (about US$150) in my account. This year and next, I won't even need to borrow seeds from our community seed bank because I have the means to purchase them, and I have also saved seeds for planting. Someone else will be able to access the seeds I would have borrowed.

"It is truly the community organization that facilitates all the activities. PDL has supported us with important knowledge, trainings, and the strengthening of our organization to help us create IGPDS. If IGPDS didn't exist, these things wouldn't be possible. IGPDS has allowed us to build relationships with and involve more people in our community. Now I see life differently because I have come to understand that I should help others insofar as I am able. Within IGPDS, we created a schedule where we all work together on each other's farms (*konbit*). There is more awareness about environmental degradation, and members of the organization encourage others to protect the environment around them.

"More people ask my opinion now, especially about agriculture, and I give them advice on how to prepare and improve their farms. Whenever this happens, I feel important. A long time ago, youth and women weren't too engaged, but now they play all roles. We have had trainings to learn about rights. I know that we have a right to eat, to have access to education and to health. Now we feel that the rights of women and children are more respected.

"Those who lead the organizations are respectful and serious, and many people trust the organization because of this. Our organization is starting to have a growing voice in the community. I am engaged in the struggle for this community to advance, and for it to offer more services. Thanks to IGPDS and the support of PDL, everyone has come to an agreement to work together. Now we are fighting together to change the living conditions of our members, and for more people to have access to seeds, and for better plans for our farms. Thanks to the access to local seeds and community loans, more people have come to our area. Now, if one day I come to complain of hunger, it will be because of my own negligence!"

management, Haiti has not had a fully functioning elected parliament or set of local government officials in place since 2010. Nonetheless, PDL and Groundswell International have provided peasant associations with trainings in civic engagement and human rights, helping them engage NGOs and international funding and development agencies to develop a common agenda for promoting healthy farming and food systems in northern Haiti.

Cost effectiveness: Bottom-up organization vs. Typical development projects

PDL's approach for strengthening peasant associations to spread agroecological farming and rural development strategies is much more cost effective and efficient, and generates more lasting impacts, than many large-scale programs of the government and international aid agencies. Strong organizational capacity and agency of peasant associations enables them to effectively lead community development processes to improve human and social capital. Also in the PDL model, farmers decide to adopt and sustain improved practices, such as soil and water conservation, because it benefits their families. This stands in contrast to many development projects, in which the only thing motivating farmers are gifts and subsidies. Finally, farmers involved in this process manage complementary local resources, such as revolving savings and credit funds, seed banks, and collective labor.

The available data reveals an impressive cost-benefit analysis for investing in peasant associations to spread agroecology and other beneficial practices. PDL's total budget over the five-year period from 2009 to 2014 was about US$3,121,000, or an average of US$624,200 per year.[6] Based on these figures, Table 3.3 shows the costs of some key benefits per year and over five years.

Table 3.3 Cost benefit analysis of PDL program

Results	Amount	Annual Cost per Result	Total Five-Year Cost per Result
Farmers learning agroecology	20,545	US$30	US$152
Farms under agroecological production	6,875	US$91	US$454
Hectares under agroecological production	687	US$909	US$4,543
Number of model farmers	11,500	US$54	US$271
Number of agricultural volunteers	646	US$966	US$4,831
Number of farmers associations	17	US$36,717	US$183,588
Numbers of members	24,580	US$25	US$127
Numbers of *gwoupman*	1,548	US$403	US$2,016
Number of blocks (communities)	203	US$3,075	US$15,374

An interesting point of comparison is the $129 million WINNER program (Watershed Initiative for National Natural Environmental Resources), launched at the same time, in 2009, by USAID.[7] WINNER has been implemented in Haiti by Chemonics International, a for-profit contractor. In 2013, Oxfam America, as a part of its work to assess aid effectiveness, produced a critical analysis of the WINNER program.

Key strategies of the WINNER program included setting up agro-supply stores to deliver inputs to farmers, and the training of farmers on best practices for production. This reflects typical agricultural development project dynamics focused on delivering external inputs and knowledge, rather than strengthening farmer agency and capacity or generating knowledge locally. Oxfam was also critical of a WINNER activity to deliver post-harvest storage bins (a valuable infrastructure) to a farmer association. The storage bins were too large for the volume of local production and too expensive to transport; additionally, adequate strategies were not developed with the peasant association to share the new resource among members. In late 2013, USAID responded to the Oxfam analysis, claiming WINNER had achieved the following:

- Increased the output of nearly 15,000 farmers, generating more than $7 million in income
- Introduced improved seeds, fertilizers, and technologies to over 17,000 farmers
- Trained 1,689 Master Farmers, who could teach others
- Increased beneficiary farmers' rice yields by 129 percent, corn yields by 368 percent, bean yields by 100 percent, and plantain yields by 21 percent (Lentfer, 2013)

One farmer-leader of the peasant association commented, "A professor showed a document telling what WINNER had done, but it meant nothing to me … The presentation might be great, but they haven't accomplished anything for farmers."

Assuming a $129 million budget, a rough cost-benefit analysis using the same approach of dividing results for each activity by the total budget of WINNER over the same period (2009–2014) would indicate:

Table 3.4 Cost benefit analysis of WINNER program

Results	Amount	Annual Cost per Result	Total Five Year Cost per Result
Farmers increasing output	15,000	US$1,720	US$8,600
Total income generated	$7,000,000 ($467/farmer)	US$3.69 (to generate $1)	US$18.43 (to generate $1)
Farmers with access to improved seeds, fertilizers, and technologies	17,000	US$1,518	US$7,588
Master Farmers trained	1,689	US$15,275	US$76,377

As with PDL, the WINNER budget was also used for other activities beyond agriculture, such as the construction of large flood and soil erosion barriers. But even if we keep the PDL budget constant, and consider only half the budget of the WINNER program for agricultural improvement activities with farmers, PDL's strategy was still nearly eight (in the case of master farmers trained) to 15 times (in the case of numbers of farmers increasing output) more efficient (USAID, 2012).[8]

Table 3.5 Comparative efficiency of pdl and WINNER programs

	Farmers	Farmer Promoters
WINNER (assuming half of US$129 million budget = US$64.5 million)	Farmers increasing output = US$4,300 five year cost	Master farmers = US$38,189 five year cost
PDL (assuming full budget)	Model Farmers = US$271 five year cost	Agricultural Volunteers = US$4,831 five year cost
Ratio	PDL is over 15 times more efficient (4,300/271)	PDL is 7.9 times more efficient (38,189/4,831)

PDL's approach also had clear positive impacts on strengthening the capacity and agency of peasant associations to lead and sustain autonomous development processes, whereas the WINNER program seems to have had the opposite effect. In addition, production improvements related to agroecology are more likely to be sustained over time, and not dependent on external inputs.

Lessons and next steps

PDL's approach has succeeded in mobilizing large numbers of small-scale farmers in rural communities, broadening the leadership base, and promoting robust, democratic decision-making and strong farmers' organizations with the capacity to improve wellbeing and agency. The peasant associations are making significant progress in scaling agroecological farming practices and other beneficial activities. Some key lessons can be drawn. First, the **strong organizational capacity of peasant associations** is inseparable from their ability to spread agroecological principles and practices, and to create an ongoing process of farmer led innovation. **Complementary strategies** to mobilize and manage local resources (seeds, grain storage, savings and credit funds, etc.) synergistically strengthen the spread of agroecology. Farmers adopting agroecological practices clearly **improved their production, income, food security, and resilience to drought and climate variability**, in comparison to farmers who did not, across both mountainsides and areas of flatland farming.

Many challenges remain. PDL and peasant associations are working to better evaluate and document the impacts of their activities. They are working to build upon initial success and develop a network of cooperative farmer enterprises to store, process, and sell healthy local food to the local population. They hope to make wider use of popular radio and other communications channels to promote agroecological production and local food consumption. More effective strategies must be found to engage with and influence large-scale rural development programs of Haiti's ministries and international development agencies.

Peasant associations are demonstrating their capacity to regenerate farms, rural communities, and livelihoods. In the face of incredible odds, these Haitian citizens are helping to build the foundation for a more resilient and beneficial future for their country. For that future to become reality, appropriate support and policies must be enacted to let them do the job.

Notes

1. Cantave Jean-Baptiste. Interview by Ben Depp, February 2014 and by Steve Brescia, March 2015.
2. Silmène Veilland. Interview with Cantave Jean-Baptiste, July 7, 2016.

3.   Cantave Jean-Baptiste. Interview by Steve Brescia, August 20, 2015.
4.   A *carreau* is a Haitian land measure equaling 3.18 acres, or 1.29 hectares.
5.   Roland Moncette. Interview by Cantave Jean-Baptiste, July 6, 2016.
6.   These cost estimates are actually high, because the PDL's program is holistic. In the table the total budget is divided for each activity, not separating out the specific budget amount by activity. In addition, this same total budget has also supported other activities not accounted for here, such as improving community health—by working with peasant associations to increase the number of latrines, water filters and purification systems, and promote health education, PDL and communities have had a major impact on preventing the spread of cholera in the area since the 2010 outbreak.
7.   The program is a part of the US Government's Feed the Future program, a US$3.5 billion global agricultural development program in response to the global food crisis of 2008-2009.
8.   In 2016, USAID stated that results include "agricultural campaigns with more than 20,000 farmers that offer packages of good practices, extension services and improved inputs."

COLOMBIA

Ibarra O

O QUITO

E C U A D O R

O Portoviejo

Quayaquil

O Riobamba

PACIFIC OCEAN

O Cuenca

PERU

Regions referenced in chapter

CHAPTER 4

Local markets, native seeds, and alliances for better food systems through agroecology in Ecuador

Pedro J. Oyarzún and Ross M. Borja

Summary: *In Ecuador, there is increasing recognition of the negative impacts of conventional agriculture and of the need to better support traditional farming practices, agroecology, and family farmers in order to improve food systems and build food sovereignty. In this chapter, representatives of the NGO EkoRural describe successes and opportunities for alliances between rural and urban people in order to build local markets, support local seed varieties, and improve biodiversity on farms.*

Rural reality in Ecuador

"In each household in our community, we have the native seeds that we have saved from our ancestors," says Elena Tenelema as she walks across her farm. Elena is from the Quichua indigenous community of Tzimbuto, in the Central Andean highlands of Ecuador, and has become a leader in managing community seed banks and in agroecological farming. "Taking care of our *Pachamama* (Mother Earth) is the most important thing. If we contaminate it with chemicals, it will be the end of our land and we won't have it in the future. Our *chacra* (smallholder family farm) is very diversified. We do not have large plantations, so we can't waste land. So for example, if I plant corn, I'll also have seven or eight other crops."[1]

For generations of indigenous people in Ecuador, farming has been a way of life. Their way of farming shares many characteristics with what is now known as agroecology. Since before colonial times, Andean farmers have exchanged goods and services, and transmitted culture and knowledge, through social relationships embedded in their strategies of production and reproduction (Barrer et al, 2004; Tapia et al, 2012). Up until the 1960s, the vast majority of people in the agricultural sector lived on small and medium-sized farms in a rural environment defined by the *hacienda* or *huasipungo* system, in which large plantation owners controlled peasant labor by allowing them to live on and cultivate small plots of marginal land (De Noni, n.d.). Since then, through two agrarian reform processes, the rural population has moved in two directions. Some rural communities gained control of large expanses of former hacienda land under collective arrangements. This land was frequently on steep mountainsides, of poor quality, and was rapidly sub-divided into small individual farms. On the other hand, a large segment of the rural population became partially dependent on wage labor in both rural and urban areas.

In the 1960s, agricultural development strategies in the Andes focused on Green Revolution technologies, followed by "modernization" in the 1980s, which emphasized strong support for export-oriented crops and agribusiness. These strategies generated an environmental, social, and economic crisis by the end of the twentieth century. Massive internal and external migration devastated traditional culture and farming practices, putting the very existence of the traditional Andean *chacras* at risk. Policies favoring the production of commercial crop varieties also reduced the genetic variety and diversity of traditional seed systems and family farms, lowered the quality of food in both rural and urban areas, and increased the concentration of production among a handful of companies in sectors like sugarcane, palm oil, and bananas.

In the fragile ecosystems where traditional family farming persists, what emerged in many cases was the hybridization of traditional practices with Western technologies. The results were disastrous. For example, using plows and disc harrows to till volcanic soils on mountainsides breaks up the soil and leads to rapid soil erosion. Farmers are left struggling to cultivate a layer of cement-like volcanic ash, known as *cangagua* (Zebrowski and Sánchez, 1996). Over half of the agricultural soil in Ecuador is seriously degraded, and the situation is particularly acute on the steep slopes of the mountainous central provinces (Fonte et al, 2012). Farmers have also eliminated fallowing practices, and have been driven to expand the agricultural frontier into fragile forest areas, furthering the degradation of soils and natural resources.

Similar to other countries, these developments also undermined food sovereignty—the local decision-making power over the production, circulation, and consumption of food (La Vía Campesina, 2011).

Traditional Andean agriculture and the evolution of family farming

In spite of these dynamics, farmers in some regions in the inter-Andean valleys have held on to their cultural knowledge and traditional food and production systems. Many elements of what is now known as agroecology can be identified in these traditional Andean farming systems (Altieri, 2011). These include: a deep knowledge of local flora (and its uses for medicine, food, and fodder); production practices that intensively use biodiversity (associations of crops within and between species, polycultures and mixed farming, tolerance of certain atypical plant populations, weeds, agroforestry, and creating earth ridges to reduce wind impact and to aerate the soil, thus reducing pests and phytopathogens, etc); terracing; the collection and application of organic fertilizers; fallowing; staggered planting schedules; dispersal of small plots on different altitudinal levels throughout the countryside; and social structures

Photo 4.1 Farmers harvesting native potato varieties, reintroduced with the support of EkoRural, Carchi, Ecuador, 2010.
Photo Credit: Ross Borja

that allow for shared local labor, as well as the circulation and exchange of complementary products among the different ecosystems of the highlands, coast, and Amazonian regions of Ecuador (Poinsot, 2004).

Many government and development actors in Ecuador continue to insist that traditional family farms are unproductive and inefficient in terms of yield/area (Benzing, 2011). Yet evidence increasingly demonstrates that it is time to reexamine this critique and to understand the multifunctional nature of agriculture. For example, research shows that family and community-based agriculture contributes between 50-70 percent of the daily food consumption of Ecuadorians, including the majority of staples such as fresh milk, rice, corn, potato, vegetables, beef, pork, and beans (Chiriboga, 2001 and 2012). Amazingly, smallholder farmers achieve this while using only 20-30 percent of the agricultural land in the country, much of it with very marginal soils. In addition, agroecological family farming has great capacity for generating employment due to its intensive use of manual labor. But land ownership remains highly concentrated in Ecuador. The Gini coefficient (which measures inequality) for land ownership has improved little between 1954 (0.86) and 2001 (0.80) (INEC et al, 2001; Castro, 2007; Hidalgo et al, 2011).[2]

Family farmers also play a crucial role in managing the biological foundation of the country's food security by using, conserving, and developing seeds (both local and improved varieties): maintaining diverse species and varieties;

dispersing plots among agroecological altitudes and ecosystems; and under-standing and using the biological indicators of climatic adaption (Poinsot, 2011; Borja et al, n.d.). Modern agriculture has much to learn from these traditional knowledge and management systems.

As has been shown from research in other countries, when compared to conventional agriculture, agroecological family farming in Ecuador is highly productive, generates multiple benefits, and has demonstrated its potential to feed the country's population (Anonymous, 2014; IAASTD, 2014; Nwanze, 2011; De Schutter, 2010).

Emerging challenges and opportunities

Since 2005, the number of organizations adopting agroecological practices in Ecuador has increased, and local market initiatives are growing. Additionally, traditional knowledge is being recovered and water, agrobiodiversity, forests, *páramos* (high, tundra ecosystems), and mangroves are being conserved, restored, and protected. Because of the need for alternatives to industrial agriculture in the face of climate change, many technicians, academics, and politicians are increasingly paying attention to smallholder farming and agroecology in search for inspiration and solutions.

However, the structural crisis in the countryside persists, without any substantial progress in overcoming the difficult economic conditions faced by rural families. To cope, many rural (and urban) people have adopted new eating habits, including more dependence on processed, industrialized foods with lower nutritional value. Food sovereignty is eroded as rural families have less control over how they farm and how they market and consume food (Ovarzún et al, 2013; Boada, 2013).

In recent decades, coalitions of rural, indigenous, and urban people's organizations have demanded an agrarian transformation. In 2008 they managed to incorporate these demands into the 2008 Ecuadorian Constitution.[3] These include support for food sovereignty; equal access to land, water, and biodiversity; the promotion of agroecology; recognition of the rights of nature; the human right to water; and the right to social participation in decision-making. The constitution has declared Ecuador a country free of genetically modified crops, and recognizes alternative practices of caring for the agroecosystem as factors that contribute to food sovereignty (Daza and Valverde, 2013). However, it has been a complex and challenging process to construct the appropriate laws, policies, and national plans to put these ideals into practice, and to ensure that family farmers and indigenous communities take part in these policy-making processes (IFOAM, 2011).

Response: EkoRural's strategy

EkoRural is an Ecuadorian NGO that strengthens endogenous (locally generated) and people-centered development processes that give a leading role to families and communities in the creation of sustainable farming and resource management strategies. We contribute to broad social change by supporting the generation of new relationships within and between rural and urban communities for the "co-production" of healthier, more democratic farming and food systems. Our methodology combines two core strategies. The first is direct support for initiatives with rural communities, including the horizontal, farmer-to-farmer diffusion of agroecological innovations. Second, we facilitate and engage in exchanges between rural community organizations, local consumer organizations, universities, local governments, and other actors, in order to spread and learn about useful strategies and practices and create new market relationships.

To create diversified and sustainable livelihoods, family farmers in the mountainous Andean region distribute their farm plots across different altitudes and micro-ecosystems. This creates a complexity that is difficult to replicate or "scale" in the classic sense of spreading a package of defined technologies. It is possible to spread key principles of farming management and to strengthen key competencies. For these reasons, EkoRural believes "scaling-up" merits a deeper discussion into how social change and innovations occur, and the roles of support organizations like ours. We seek to strengthen our own, but more importantly farmers', understanding of the micro-ecosystems as well as the wider food system, and the different skills and strategies relevant to each.

In terms of scaling or promoting wider systems change—rather than focusing on strategies to expand our work directly to a large and growing number of communities—we work to foster viable models that serve as inspiring examples, and to share these through wider networks. These networks include the Agroecology Collective, COPISA (Intercultural Council on Food Sovereignty), and MESSE (Movement for the Social and Solidarity Economy of Ecuador), as well as international networks such as PROLINNOVA (Promoting Local Innovation), several communities of practice, and Groundswell International.

In rural communities, we find family farms that are regenerating, at equilibrium, or degenerating. This heterogeneity is a starting point to develop strategies for the spread and intensification of agroecology, empowering family farmers to develop balanced farming systems and to work for food sovereignty. Production must meet the needs of the family farm in regenerative ways, and excess production should go first to local markets through more direct connections between the farmers and consumers. This reciprocal relationship between the countryside and cities is central to agroecology.

For these reasons, we have increasingly carried out our development and research initiatives within the framework of *Healthy Local Food Systems*. Our hypothesis is that healthy local food systems (family farms using local resources and knowledge, that have relationships with consumers, and that protect the health of people and the ecosystem and contribute to sustainable livelihoods) generate greater wellbeing and are more resilient to social and climactic changes than are conventional food systems (based on industrial agriculture that generates negative effects on health, the environment, the economy, and culture).

EkoRural manages two programs located in the Northern and Central Highlands and works directly with ten communities and 500 family farmers, and indirectly reaches about 2,000 farmers. We involve family farmers in processes of action-learning, knowledge development, and sharing through farmer-to-farmer exchanges. In this way, we support technological change and strengthen local leadership and organizations for broader social change. Entry points of interest for family farmers are soils, seeds, water, and strengthening of rural-urban relationships and local markets. Some of the key activities include:

- **Soils:** systems based on cover crops, green manures, and limited tillage that reduce soil-degradation and increase fertility
- **Water:** water harvesting and efficient micro-irrigation and water use
- **Seeds:** strengthening the capacities of farmers and local organizations to conserve, use, and manage agro-biodiversity, based on the recovery of Andean food species and strengthening local seed systems
- **Local markets:** strengthening rural-urban relations

Photo 4.2 Varieties from a local community-managed seed bank, Tzimbuto, Ecuador. *Photo Credit:* Steve Brescia

Strengthening community management of seed systems and agrobiodiversity

Discovery-based Learning

By strengthening their management of agro-biodiversity, communities can become more resilient to climate change. We support a discovery-based learning process of the complex relationships between families' livelihoods and their management of seeds and other biological resources. We generally start with a participatory identification of available biological resources on farms and in the community, as well as the wide range of practices used to manage those resources. We then work with communities to promote more diverse, resilient, and productive farming systems through activities such as: improved on-farm management of biodiversity; participatory plant breeding; seed banks; and strengthening of community-based organizations to manage these processes.

Farmer innovation and management

Some of the key innovations that farmers test and spread are: the introduction of new species and varieties; sowing schedules; crop rotation; production of compost; recycling of organic matter; and other organic soil inputs. In addition, we support the recovery and reintroduction of native varieties of potato and other Andean tubers, thereby increasing genetic diversity and diet on family farms. In the last five years, we have helped to recover and promote the use of dozens of varieties of species that had fallen into disuse (such as *machuas*, *ocas*, *mellocos*, *jícama*, local beans, and native potatoes, among others) and put them into circulation in communities. We have returned an important part of the Ecuadorian potato collection to the hands of small-scale farmers and have participated in the distribution and testing of certain varieties high in zinc and iron. To aid in the conservation of potato varieties, we have slowly introduced the ideas of precocity and disease-resistance.

💬 **Farmer testimony**

Juan Simón Guambo, farmer and leader from Flores Parish, Riobamba Cantón, Chimborazo Province[4]

"Climate change sometimes makes the weather too hot and sometimes too cold, damaging my crops. So we have been planting native species of plants around the chacras, such as oca, mashua, ulluco, native potatoes and maize, provided by the provincial government and EkoRural, and we also learned about agrobiodiversity and soil management. I am very proud of the biodiversity I now have in my chacras, of the fact that I can share new experiences such as how to propagate plants and how to recover microclimates with windbreaks around the chacra. I want to see my entire family learn this expertise and continue applying it in their chacras instead of migrating from the region to learn things that have nothing to do with our culture and customs."

On the farm level, we have worked with farmers to integrate the concepts of "ecological confusion,"[5] heterogeneity, and multiple uses of space. This has resulted in improved nutrient use and mobility, pest control, and the continuous flow of products for family consumption and sale in local markets. Local consumption of their own products helps to generate a self-sustaining loop of production and consumption. This closed-circuit system, which also includes large quantities of organic fertilizer produced from families' own livestock, contributes significantly towards their greater autonomy (Marsh, 2017).

Community revolving seed banks

Women farmers play a key role in producing and circulating the seeds of local food species through revolving seed banks. To sustain and grow the seed banks, farmers return two units of seeds for every unit borrowed, which can in turn be lent to more farmers. This creates a redistribution method that circulates quality, local seed, while also generating a rotating community development fund. For example, in the community of Chirinche Bajo, in the province of Cotopaxi, a community seed bank was initiated several years ago with 25 pounds of potato seed. Members have now produced over 110,230 pounds of potatoes that they have sold, traded, re-planted, and consumed. Each seed bank has its own dynamics depending on the seeds they manage.

The diversification and complexity of crop rotations has resulted in a notable increase in functional biodiversity, reaching an average of 30-40 species in some plots (for example, crops, fruit trees, medicinal plants, etc.). As one community organizer commented: "Our community has been able to change its production into a system that uses crop rotation, diversification of crops, and organic fertilizer, which as a whole has contributed towards diversifying both our harvests and our diets" (Marsh, 2011). At least 50 varieties of Andean root and tuber crops (mainly potatoes, *mellocos, ocas,* and *mashuas*), have been introduced by identifying and recovering varieties still produced by some farmers in the region. Community-led reproduction and dissemination, and participatory plant breeding has improved the varieties. For example, the I-Libertad potato was officially released and has been widely disseminated on 500 farms in ten communities via direct and farmer-to-farmer support, allowing EkoRural to catalyze the spread of quality local seed varieties and agroecological practices. Neighboring communities, as well as other organizations, are now learning of these innovations and are establishing their own seed banks, using their own systems of exchange and control. New opportunities have been created for women in managing and selling these genetic resources. This process has also deepened our understanding of community management of genetic resources in the context of a changing climate and evolving farming practices in the Andes.

> ⌾ **Farmer Testimony**
>
> *Elena Tenelema, Tzimbuto, Central Ecuadorian Highlands[6]*
>
> "Thirty years ago all we cultivated here was corn and some beans. With the support of EkoRural, we have recovered seeds and plants that we had stopped producing. Now we learn and share with other communities about what seeds and plants they still have. Then, with a small group of ten or so of us, we start to test and reproduce the seeds on our own farms and community plots. For every pound of seeds we receive, we commit to passing on two pounds to other farm families—while keeping enough to continue producing ourselves. The 24 families currently involved have all developed diverse agriculture plots with many new food plants. A wider group of 52 families has become interested, and we are supporting them to learn about what we are doing, obtain seeds, and develop their own diverse plots. We are working to reach all of the 150 families in Tzimbuto to ensure that they have access to all of these seeds they need to have diverse farms and healthy food to eat."

Strengthening local food markets and urban-rural linkages

If people don't eat healthy local foods, then quality local seeds and community biodiversity, key to agroecological farming, will disappear. So over the last five to ten years we have promoted a process for forging direct, win-win relationships between farmers and urban consumer organizations to strengthen local food systems. In practice, this has resulted in empowering farmers, increasing their incomes, and strengthening their ability to negotiate with buyers. Consumers gain access to healthy, local food at a lower cost—while supporting agroecological farming. Producers from several communities have joined the *Canastas Comunitarias* movement (Community Baskets, a model similar to "Community Supported Agriculture" or CSA agreements) and started direct sales and agroecological farmers' markets and fairs. The Canastas and alternative food networks foster more personal, beneficial, and transparent relationships between urban and rural organizations; raise public awareness; and provide opportunities to address issues such as gender relations and appropriate policies for food security, rural investment, and biodiversity. In the words of farmer Lilian Rocío Quingaluisa from the province of Cotopaxi: "Engaging directly with urban citizens is great for us as women farmers. It means we have better income, we do not have to work on other people's land, we are more independent, and we can spend more time with our families and animals".[7] Another farmer, Elena Tenelema, adds: "The baskets eliminate abuse by intermediaries. Second, they give us a guaranteed income, which we can use to improve our health, for education, or to buy animals. People in town get to know and eat our products. That is one of the most important things that we are fighting for as indigenous farmers."

There is growing recognition of these kinds of promising local market initiatives in the political sphere in Ecuador, and the constitution recognizes them under the framework of Social and Solidarity Economics. But fostering direct and reciprocal food systems is not an easy task, especially in the face of industrialized agriculture and food distribution, and much work remains to be done.

Photo 4.3 Canasta comunitaria in Riobamba.
Photo Credit: Steve Brescia

The "250,000 Families!" Campaign: Farmers and Consumer Citizens as a Force for Change*

In 2005, Ecuador's rural-based agroecology movements got together with an urban-based wholesale purchasing group, the *Canastas Comunitarias* (Community Food Baskets) to exchange experiences. One conclusion was that, in its enthusiasm about farming practices, the agroecology movement had inadvertently isolated rural producers from urban-based consumers. The resulting *Colectivo Agroecológico* shifted its attention from "good agronomy" to "good food"—a more holistic platform, which seamlessly linked rural and urban people around a common cause. Their rally call became "food sovereignty:" food for the people, by the people, of the people.

The *Colectivo* played a central role in influencing Ecuador's groundbreaking 2008 Constitution and the subsequent national policy transition from food security (understood as merely meeting peoples' basic needs) to food sovereignty (an emancipatory force for democratic change). Following a decade of advocating for food sovereignty, the *Colectivo* concluded that the dominant food system that it so fervently criticized—what may be the single largest industry on the planet (estimated to be worth over US$1.3 trillion per year in places such as the United States and about 10 billion per year in Ecuador)—had become so influential in national politics that it was no longer realistic to expect government representatives to be able to correct things on their own. Ultimately, people operating both individually and collectively in the families, neighborhoods, and social networks that cross urban and rural environments, must wrest control over their food territories and their futures. This is the vision of "consumer-citizens:" they are actively informed, take a position, and act in their own better interests.

In October 2014 the *Colectivo* launched its "250,000 Families!" campaign—a five-year project to recruit 5% of Ecuador's population to take charge of making food sovereignty a reality. Through shifting about half of the present food and drink purchases of this population, these consumer-citizens would invest about US$300 million per year in healthy local food production: more than the total spent on international cooperation for

---

(Continued)

> **Box** Continued
>
> agriculture and health in Ecuador. In order to become part of the 250,000 campaign, a family must address two questions: what does "responsible consumption" mean for me, and how does my family (business or community) practice it?
>
> *Adapted from: Sherwood, Stephen and Caely Cane. "250,000 Families! Reconnecting urban and rural people for healthier, more sustainable living." Urban Agriculture Magazine, number 29, May 2015.

Lessons and recommendations

While there is reason for serious concern regarding family farmers in Ecuador, there is also reason for optimism. The wealth of experiences in family farming, the wide number of actors involved in the agroecological movement, and growing alliances around healthy local food systems can generate significant and positive changes in practice and policy. But they also face opposition.

We need a new paradigm for agricultural development. This begins by recognizing the multi-functionality of agriculture and abandoning the agroindustry's narrow focus on export-oriented production based on external inputs. The dogmatic application of these uniformly prescribed solutions needs to be replaced with a focus on strengthening the competencies of family farmers and rural organizations so that they can innovate, applying their knowledge, skills, and values to their unique, local contexts. Evaluating the progress of such programs should involve flexible frameworks that take into account the motivations and values of local people.

Government policy in Ecuador is contradictory and incoherent. On one hand, current development policies support stronger public institutions, some degree of redistribution of income, and increased access to public services— including for rural citizens. In the agricultural sector, the Constitution and some laws affirm food sovereignty and agroecology. But in practice, current policies support agroindustry and subsidized inputs, rather than landless farmers and smallholders with the means to become more productive, sustainable family farmers. At the moment, the government has almost abandoned programs for land and water redistribution and is promoting subsidized chemical fertilizers and certified seeds, fostering increased dependency on commercial services. The government is attempting to modify the Constitution to allow the entry of genetically engineered crops into Ecuador. This constitutes an attack on agrobiodiversity and on the health of the people and ecosystems (Anonymous, 2014). Ecuador expects to sign a new trade agreement with the European Union, and it is unclear what impact this will have on agriculture.

Rural development and research programs should support new ways of producing and distributing food. This includes raising awareness and taking action to strengthen democratic, healthy local food systems. In addition to our work with farmers and their organizations, we must create productive dialogue and linkages across public institutions, civil society, NGOs, universities, research institutes, and rural and urban communities. This includes collaborating with influential urban networks and consumers' organizations. We need to be constantly aware of innovations in urban-rural relationships, including peri-urban and urban agriculture. As Pacho Gangotena, farmer and agroecologist says, "I believe that social change in agriculture will not come from above, from the governments. It will come from the thousands and millions of small farming families that are beginning to transform the entire productive spectrum … We are a tsunami that is on its way."[8]

Notes

1. Elena Tenelema. Interview with EkoRural, 2012.
2. Additionally, the III National Agricultural Census (2000) showed that there are 600,000 families that are divided into farms of 1.5 hectares or less, while there are 1,300 proprietors with plots over 500 hectares, controlling 1.8 million hectares. From a total of 841,000 UPAs (*Unidades Productivas Agrícolas* or Productive Agricultural Units), 740,000 correspond to the sector of family agriculture.
3. As specified, for example, in: the Ley Orgánica del Régimen de Soberanía Alimentaria of 2009; the Plan Nacional para el Buen Vivir 2009-2013; the Plan Nacional para el Buen Vivir 2013-2017; the Ley Orgánica de Economía Popular y Solidaria y del Sector Financiero Popular y Solidario.
4. Juan Simon Tambo. Interview with EkoRural, 2015.
5. Used to promote on farm genetic diversity to help control and manage pests and diseases.
6. Elena Tenelema. Interview with EkoRural, 2012.
7. Lilian Rocio. Interview with EkoRural, 2014.
8. Pacho Gangotena. Interview with Ben Depp of EkoRural, June, 2014.

OREGON

IDAHO

PACIFIC
OCEAN

NEVADA

UTAH

C A L I F O R N I A

SACRAMENTO

San Francisco

Santa Cruz

Fresno

ARIZONA

Los Angeles

San Diego

MEXICO

Regions referenced in chapter

CHAPTER 5

Agroecology and food system change: A case study of strawberries in California, The United States of America

Steve Gliessman

Summary: *Since the 1980s, I have been working with US farmer Jim Cochran, experimenting with agroecological ways to grow strawberries in California while building alternative food networks. In 30 years, the revenue of organic farming in these counties increased by 2,000 percent. We learned important lessons for how to scale out and scale up the agroecological transition by combining techniques (such as diversification, rotation, and multiple cropping), building on previous results, sharing lessons with other farmers, and connecting with consumers to create new markets. The broad and continuous integration of research, practice and social change was fundamental in this process.*

The central coast of California in the United States, with its mild Mediterranean climate, is probably the most important strawberry growing region in the world. On approximately 15,366 acres, the Californian counties of Monterey and Santa Cruz together produced more than US$976 million worth of strawberries in 2012, about half of the total California crop.

While most strawberry production here is highly dependent on expensive, energy-intensive, and often environmentally harmful off-farm inputs, its acreage of organically grown strawberries has increased sevenfold since 1997. The partnership around agroecology I developed over decades with a farmer has made significant contributions to this change. It started in the early 1980s, with the rise of consumer interest in organic food as a result of the health and environmental hazards caused by pesticides.

The present system of industrial monoculture strawberry production in California can be traced back to the early 1960s, when the soil fumigant methyl bromide (MeBr) was introduced. Researchers and farm extension agents promoted MeBr as the way to overcome the rapid buildup of soil-borne plant pathogens that did not allow for long-term, continual production of strawberries on the same piece of land. Until that time, growers treated strawberries as a perennial crop, where plants were kept in the ground for two to four years, after which each field required rotation out of strawberries for several years.

Photo 5.1 Professor Steve Gliessman (left) and farmer Jim Cochran (right) have collaborated to create a transition to agroecological strawberry production in California.
Photo Credit: Manolis Kabourakis

However, the use of methyl bromide from the 1960s onwards allowed growers to manage strawberries as an annual crop by planting new plants year after year on the same piece of land. In that system, strawberry plants are removed each year following the end of the growing season, after which the soil is cultivated and fumigated before being replanted with new plants for the next season. Intensive systems of drip irrigation, plastic mulch, and soil manipulation are required. Breeding programs that had been in progress before MeBr to develop disease resistant strawberry varieties were abandoned and the germplasm was lost as breeders focused on maximizing the yields of fruit for shipping to rapidly expanding national and international markets. For such a high value crop, and with the cost per acre of production easily exceeding US$25,000, MeBr removed most of the risk involved in growing strawberries.

In the early 1980s, as interest in organic food became a potential market force in agriculture and issues of pesticide safety and environmental quality came to the fore, farmers began to move away from the use of MeBr and developed new practices.

In this context, for over 30 years, I have built a unique relationship with a farmer named Jim Cochran and his Swanton Berry Farm in Davenport,

California, on the northern rim of Monterey Bay, where many strawberries are grown. Our relationship has allowed us to carry out a multifaceted research collaboration centered on studying the process of converting a conventional strawberry production system into a more sustainable organic agroecosystem. We have done this using agroecology as our guiding foundation, and our journey toward sustainability has taken both of us from his field, to the market, and to the table of the consumers who have supported him.

This relationship shows that even systems strongly invested in industrial/ conventional practices can be changed; it also exemplifies the difficulties and barriers inherent in converting or transitioning to a new food system model. Moreover, from our collaboration, and as our thinking evolved, grounded theory emerged about the "levels" of the transition process to sustainability. Our experience provides useful insight into how to scale out and scale up the agroecological transition process, as well as insight into the changing role of science in this transition.

How it started

When we planted our first plots together on his original three-acre farm in 1986, we were told by conventional growers and the local strawberry agriculture extension specialist that it was impossible to successfully grow organic strawberries.

But as an agroecologist in the early stages of developing what was probably the first formal academic program in agroecology in the world, at the University of California at Santa Cruz, I was convinced that an ecological approach to agriculture could solve the problems that we would confront in the transition to organic management. Jim, on the other hand, was a beginning farmer in the process of getting organic certification after several years of working with organizations that were following the conventional MeBr model of strawberry production. His direct exposure in the past to MeBr, as well as other toxic chemicals, convinced him that there had to be another way.

It was serendipitous that his first plantings were just over the fence dividing his field from the home I was living in at the time. Over that fence, our talks about transition led to the first side-by-side comparative trial of organic strawberries. Our plots were on his land, using his varieties and practices, his workers, and many of his resources. Our research was funded by the newly established University of California's Sustainable Agriculture Research and Education Program (UCSAREP). This program had been mandated by the California legislature in 1984, and required that after many years of neglect from the Land Grant system, the University put resources into meeting the needs of small farmers, farm workers, and alternative farming systems that included organic farming. Without this program and the funding it provided, our conversion study may never have begun. A relationship began that continues to grow and evolve up to the present.

Photo 5.2 Industrial/conventional strawberry field fumigated with methyl bromide near Watsonville, California. Vaporized MeBr is held under the plastic for several days. Conversion to organic management involves replacing this very toxic and expensive chemical with a variety of alternative practices.
Photo Credit: Steve Gliessman

The transition process

The year-by-year evolution of the relationship that was built, the projects and activities that were carried out, and the "levels" for which each step occurred in the transition process, is outlined in Table 5.1. The five levels of transition are presented in the recent third edition of my agroecology textbook (Gliessman, 2015), and are a useful way for understanding how to scale out or scale up the agroecological transition process. A summary of these levels is presented in Table 5.2, organized according to the three different "aspects" of agroecology: research, farmer collaboration, and social change.

Level 1 conversion: Input reduction

My first efforts related to conversion, carried out before Jim and I connected over the fence, were focused mostly on finding more effective ways of controlling pests and diseases so that inputs could be reduced and their environmental impacts lessened. Many of the conventional chemicals used in strawberries were being removed from use due to increasing evidence of their negative impacts. But these regulations were beginning to limit options for farmers. So we tested, for example, different miticides for control of the common pest, two-spotted spider mite (*Tetranychus urticae*), with the goal of overcoming the problems

Table 5.1 Chronology of the transition towards food system change*

Date	Activity or milestone	Conversion Level
1986	Contact with first farmer in transition	Level 1 to Level 2
1987–90	On-farm collaborative comparative conversion study	Level 2
1990	First conversion publication, *Calif. Agriculture* 44:4-7	Level 2
1990–95	Refinement of organic management	Level 2
1995–99	Rotations and crop diversification	Initial level 3
1996	Second conversion publication, *Calif. Agriculture* 50:24-31	Level 2
1997–99	First alternatives to MeBr research projects	Level 2
1998	BASIS (Biological Agriculture Systems in Strawberries) work group established to disseminate research findings	Levels 2 & 3
1999	Soil health/crop rotation study initiated	Levels 2 & 3
2000–06	Strawberry agroecosystem health study	Levels 2 & 3
2002–03	Pathogen study, funded by NASGA (North American Strawberry Growers Association)	Levels 2 & 3
2001–05	Poster/oral presentations at American Society of Agronomy meetings	Level 3
2003–06	Alfalfa trap crop project	Level 3
2004	Organic Strawberry Production short course organized by the Community Agroecology Network in Santa Cruz and the Universidad Autónoma de Chapingo in Huatusco, Veracruz, Mexico	Levels 2 & 3
2004–08	USDA-Organic Research and Extension Initiative project: Integrated network for organic vegetable and strawberry production	Levels 2, 3 & 4
2004-present	Partner grower establishes an on-farm farm stand, including value-added products such as pies, shortcake, and jams, as a complement to his farmers' market and direct sales	Level 4
2005–06	Local organic strawberries in UC Santa Cruz dining halls along with other organic produce	Level 4
2006	California Strawberry Commission and NASGA fund organic rotation system research	Level 3
2007-present	Research on alternatives to MeBr fumigation with anaerobic soil disinfestation (ASD) to shorten rotation period	Level 2 & 3
2011	USDA-Organic Research and Extension Initiative project: Support for expanded ASD research on local farms	Level 2 & 3
2014	Crop rotation and biofumigation study published, *Agroecology and Sustainable Food Systems* 38(5): 603–631	Level 2 & 3
2014	Food Justice Certification awarded to partner grower	Level 5

*Much of the early work was carried out before I retired from the university in 2012, with the collaboration of what was called the Agroecology Research Group at the University of California, Santa Cruz.

Table 5.2 The levels of transition and the integration of the three components of agroecology needed for the transformation to a sustainable world food system

Level	Scale	Role of Agroecology's Three Aspects		
		Ecological Research	Farmer Practice and Collaboration	Social Change
1 Increase efficiency of industrial practices	Farm	**Primary**	**Important** Lowers costs and lessens environmental impacts	**Minor**
2 Substitute alternative practices and inputs	Farm	**Primary**	**Important** Supports shift to alternative practices	**Minor**
3 Redesign whole agroecosystems	Farm, region	**Primary** Develops indicators of sustainability	**Important** Builds true sustainability at the farm scale	**Important** Builds enterprise viability and societal support
4 Re-establish connection between growers and eaters; develop alternative food networks	Local, regional, national	**Supportive** Interdisciplinary research to provide evidence for need for change and viability of alternatives	**Important** Forms direct and supportive relationships	**Primary** Economies restructured; values and behaviors changed
5 Rebuild the global food system so that it is sustainable and equitable for all	Global	**Supportive** Trans-disciplinary research to promote the change process and monitor sustainability	**Important** Offers the practical basis for the paradigm shift	**Primary** World systems fundamentally transformed

Source: Adapted from Gliessman 2015

of evolving mite resistance to the pesticides, negative impacts on non-target organisms, pollution of ground water, persistent residues on harvested berries, and health impacts for farm workers (Sances et al, 1982). Controlling weeds and slowing soil erosion with winter cover crops planted in small windows of opportunity between strawberry planting cycles was another research focus.

Level 2 conversion: Input substitution

In 1987, an existing partnership between a recently formed agroecology research group at UC Santa Cruz and Jim Cochran became a comparative strawberry conversion research project.

For three years, Jim was growing strawberries in plots using conventional inputs and management side-by-side with strawberries grown under organic

Photo 5.3 A side-by-side study of the conversion of conventionally grown strawberries to organic management. In this level-2 study, more sustainable inputs are substituted for their conventional equivalents. Taken at Swanton Berry Farm, Davenport, California from 1986–1989.
Photo Credit: Steve Gliessman

management. In the organic plots, each conventional input or practice was substituted with an organic equivalent. For example, rather than control the two-spotted spider mite with a miticide, beneficial predator mites (*Phytoseiulis persimilis*) were released into the organic plots. Over the three-year conversion period population levels of the two-spot were monitored, releases of the predator carried out, and responses quantified. By the end of the third year, ideal rates and release amounts for the predator—now the norm for the industry—had been worked out (Gliessman et al, 1996).

However, the agroecosystem was still basically a monoculture of strawberries, and problems with disease increased. After the three-year comparison study, our research group continued to observe changes and Jim, as the farmer, continued to make adjustments in his input use and practices. This was especially true in regard to soil-borne diseases. After a few years of organic management, diseases such as *Verticillium dahliae*, a source of root rot, began to occur with greater frequency. The first response was to intensify research on input substitution. Initial experiments with mustard biofumigation took place, adjustments in organic fertility management occurred, and mycorrhizal soil inoculants were tested.

We began further research to substitute for MeBr fumigation with a practice called anaerobic soil disinfestation (ASD). This approach incorporates different sources of organic matter, from broccoli crop residue to mustard seed cake, into the soil, floods the soil with water, then covers the soil with an impermeable plastic tarp. The combination of anaerobic conditions and breakdown products of the organic matter fulfill the same function as MeBr, but with materials accepted by organic certification standards (Shennan, 2010). The big question was whether this substitution would continue to allow monoculture organic strawberries to be produced, or if creative ways could be found to strengthen the strawberry production system through diversification and redesign of the system.

Level 3 conversion: Redesign

It was at this point in the early 1990s that a whole-system approach began to come into play. Based on the notion that ecosystem stability comes about through the dynamic interaction of all the component parts of the system, our researchers worked with Jim to come up with ways to design resistance to the problems created by the monoculture system. Jim realized he needed to partially return to the traditional practice of crop rotations that had been used before the appearance of MeBr (see Box 5.2). Our researchers used their knowledge of ecological interactions to redesign the strawberry agroecosystem so that diversity and complexity could help make the rotations more effective, and in some cases, shorter. The testing of these ideas has resulted in considerable progress. For example, we designed the crop rotations using mustard cover crops to test them for their ability to allelopathically reduce weeds and diseases through the release of their toxic natural compounds. Broccoli was shown to be very important as a rotation crop since it is not a host for the *V. dahliae* disease organism, and broccoli residues incorporated into the soil release biofumigants that reduce the presence of disease organisms (Muramoto et al, 2005; Muramoto et al, 2014). Other crops that are not hosts for the disease have also been successfully used in rotation with strawberries, such as spinach, peas, and artichokes. It took more research to choose the right species and achieve the best impact and understand the ecology of the interactions.

   Rather than rely on biopesticides, which still have to be purchased outside the system and released, we incorporated natural control agents into the system, keeping them present and active on a continuous basis. For example, we tested the idea that refugia for the *P. persimilis* predator mite could be provided, either on remnant strawberry plants or trap-crop rows around the fields. Perhaps the most novel redesign idea was the introduction of rows of alfalfa into the strawberry fields (see Box 5.1). Some of the changes at this level came from new agroecological research, and others were based on "re-learning" some of the practices used for strawberry production before the 1960s.

Level 4 conversion: Alternative food networks

Consumers have been a very important force in the transition of Jim's strawberry agroecosystem to a more sustainable design and management. By responding to consumer demand for organic produce, allowing organic

Box 5.1 Alfalfa as trap crop for bugs

One innovative and successful aspect of our redesign of the farm was the use of alfalfa as a trap crop for the western tarnished plant bug (*L. hesperus*). The pest can cause serious deformation of the strawberry fruit, and because it is a generalist pest, it is very difficult to control through input substitution. By replacing every 25th row in a strawberry field with a row of alfalfa (approximately three percent of the field), and then concentrating control strategies on that row (vacuuming, biopesticide application, predator or parasitoid releases, etc.), it was possible to reduce Lygus (*L. hesperus*) damage to acceptable levels (Swezey et al, 2013). The ability of these alfalfa rows to also function as reservoirs of beneficial insects for better natural pest control has been tested as well, with field sampling showing an abundance of natural enemies occurring in the alfalfa strips. A selective endoparasitoid (*P. relictus*) from Spain has been successfully introduced into the strips where it now breeds and helps in biological control by parasitizing nymphs of the western tarnished plant bug (ibid).

Photo 5.4 Gliessman

farming to become increasingly important, Jim sold directly to consumers through farmers' markets, a farm stand with processed products such as pies and jams, on-farm strawberry picking, direct delivery to stores and restaurants, and other sales to consumers, businesses, or organizations that showed solidarity with Jim's transition efforts. In one example, students at UC Santa Cruz convinced the campus dining service managers to begin integrating local, organic, and fair-trade items—including Jim's organic strawberries—into the meal service. The creation of these new markets allowed Jim to build direct relationships with his clients and capture a larger percentage of the sales price.

Jim connected not only with consumers but also with other producers, extending the results of the transition far beyond his farm. In the early days of

Photo 5.5 Strawberries in rotation with other crops, and surrounded by natural vegetation. This agroecosystem uses the redesign principles of level 3, but also requires connections to consumers at level 4, and a change in values and knowledge at level 5. Swanton Berry Farm, Davenport, CA.
Photo Credit: Steve Gliessman

our collaboration, we held farmer field days at his farm to showcase both our research findings and the farming practices he was developing (see Box 5.2). And we shared our insights in other ways. Over the years, we have published research results; participated in a variety of workshops, conferences, and short courses on organic strawberry production; and used Jim's farm as a place to continuously link research and practice. We even helped design, present, and publish the results of an organic strawberry workshop (Koike, 2012), although our calls for diversification have largely gone unheeded. It will take much more research on the complex process of redesigning strawberry production systems to convince farmers to take the risk of moving beyond input substitution and monoculture management. The continued growth in demand for organic strawberries by consumers is an important incentive for this to happen.

Level 5 conversion: Rebuilding the food system

Our partnership has brought about immense changes, as can be seen in Table 5.3. Despite these positive trends, several sustainability challenges are connected with this dramatic growth in strawberry production. For example, we observed soil erosion and nutrient leaching where organic strawberries are planted over a large area, as well as groundwater depletion and saltwater intrusion into aquifers. What might be called "level-5 thinking" should

---

Box 5.2 The transition process in the words of farmer Jim Cochran

"The ranch I took over in the early 80s was half planted in artichokes and half planted in Brussels sprouts. I noticed that the strawberries I planted in the Brussels sprout half were doing much better than the plants in the artichoke half. So I remembered something about crop rotations that I had read years ago. At that time there was no information available about crop rotations. If I went to the farm advisor for help, he would say: 'Jim, you are crazy, the solution to that is to fumigate—it works like a charm.' When I told him I didn't want to do it that way, he would say he had nothing else to offer.

"But then, Steve told me there is a strong history of scientific analysis of rotations, which had become lost knowledge over the last 50 or 60 years, when the heavy use of chemicals became popular. Steve set up trials on my land and started looking at particular crop rotations. He eventually found evidence that it was effective and that it wouldn't be necessary to use chemicals anymore. This is how our collaboration started.

"So when Steve came, he really solidified my path, because I was sort of flying blind. I didn't write down my rotation schedule, I didn't write down my yield-per-block. I just sort of observed that stuff. He provided the scientific matrix in which to put the information that I was starting to collect. Importantly, Steve and I had a similar outlook. The idea was to study the system that I was developing and to add a scientific underpinning to it. We would develop an alternative methodology to growing strawberries, mainly organically. There was none of that at the time.

"We decided to do a public plot that was open to other farmers to come and visit. This was in the late 80s. The plot was half organically managed and half chemical. We held a series of public meetings and various groups of farmers and researchers came to visit. That is when people began to see that it would be possible to grow strawberries organically. From that point on, more and more farmers started to experiment with growing organic strawberries on their own."

Source: Interview for *Farming Matters* by Jessica Milgroom, ILEIA, March 2016

Table 5.3 Changes in organic strawberry production in California, 1997 to 2011[*]

Year	Area in organic production (acres[**])	Gross declared value (US$ in Millions)	Number of organic producers
1997	134	n/a	n/a
1998	244	2.5	82
1999	805	8.7	99
2000	545	9.7	119
2001	756	9.3	113
2002	1,278	12.5	105
2003	1,290	24.6	99
2004	1,382	28.4	n/a
2005–2010	n/a	n/a	n/a
2011	1,638	63.5	95

[*]Data from CDFA available only for 1997-2004; most recent data only available through 2011 from USDA. Source: California Department of Food and Agriculture, California Organic Program (www.cdfa.ca.gov/is/i_%26_c/organic.html); United States Department of Agriculture, Department of Agricultural Statistics (http://usda01.library.cornell.edu/usda/current/Organic Production/OrganicProduction-10-04-2012).

[**]Acreage may tend to be an over-estimate since it may also include fallow or unplanted land set aside for future plantings.

include the consideration of such issues, as part of a concern for the health of the entire system. As can be seen in Table 5.3, the number of organic strawberry producers has declined since 2000, even as the acreage planted has increased. These trends continue to the present.

"Level-5 thinking" must also include more complex social issues such as labor rights and food justice. Since organic strawberries usually require more labor, they have the potential to provide excellent job opportunities. But worker health, safety, and pay equity must become the norm. Jim's Swanton Berry Farm is one of the only organic strawberry growing operations willing to sign a contract with the United Farm Workers (UFW) union as early as in 1998, guaranteeing wage, health, and vacation benefits. In 2013, Jim's farm became one of the first of two to achieve what is called Food Justice Certification, because of the ways he has integrated social justice into his farming practices and his relationships with his workers. His whole-system approach to farming is an important example of steps that can be taken to rebuild the food system. This entails the next important step needed for researchers: to move beyond Level 2 and 3, and link their work to the more transformative food system changes.

The results

Jim's success became an incentive for other local growers to begin transitioning their farms (see Box 5.2), especially using Level 2 substitution in order to gain organic certification. In the two Central Coast counties of California, there were a total of 35,630 organic-certified acres in 2012, more than seven times the organic acreage recorded in 1997. The total farm gate revenue from organic farming in these counties was US$247.7 million in 2012, representing a dramatic increase of more than 2000 percent from 1997 (Monterey County Agricultural Commissioner, 2013; Santa Cruz County Agricultural Commissioner, 2012). A parallel increase in organic strawberry production occurred during this same time period, as can be seen in Table 5.3.

Lessons learned and next steps

When Jim first decided to transition to agroecological farming, everyone told him that it was not possible to do it successfully. And when we joined forces in 1986, we were considered to be too radical in our thinking, if not actually crazy. But in fact, one of the most valuable parts of the collaboration has been having a friend with the same line of thinking. It really was a two-way co-creation process, with research results being presented to Jim, discussions back and forth about possible changes in the farming system and practices, bringing in research ideas from other projects, sharing them, and coming up with possible ways to put them to work on the farm. We helped to keep each other going over 30 years of challenges.

Building a close relationship between researchers and farmers that can grow, evolve, and persist is not easy. Building this relationship took time, trust, flexibility, and a willingness to share knowledge, values, and belief systems. Such a participatory and action-oriented relationship is an essential component of the way agroecology must operate in order to promote either the scaling out to other farmers, or scaling up in the food system to promote real change.

In many ways, a commitment to food systems thinking was needed from the beginning. In our particular case, the social change dimensions of agroecology were present when we first planted our comparative trials in 1986. They guided our interaction and research trajectory, and influenced Jim's own development and farming approach. We have had to constantly be on the lookout for cooptation of agroecology, be it by the large-scale, vertically integrated and market-oriented strawberry industry, or by conventional agricultural research universities or institutions. For example, current research on alternatives to MeBr, which is scheduled to finally be fully phased-out in 2017, is focused primarily on finding a replacement for the toxic fumigant, be it either another chemical or an organically acceptable practice, so that strawberries can be continuously cultivated. There is very little emphasis on redesigning the monoculture strawberry system with diversification, rotations, or multiple cropping, as Jim has chosen to do.

Looking into the future, it will be a challenge for farmers in California to adapt to the continuing drought crisis. The five levels of transition can serve as guidance in this process. Agroecological practices around input reduction and substitution (Levels 1 and 2) will reduce the need for intensive irrigation, and Level 3 redesigns will require less intensive farming practices. However, this will also most likely imply lower yields per acre, which is why the creation of new, direct markets with better prices for producers (Level 4) will be crucial. And Level 5 thinking will need to come to the fore as farmers realize that water is limited, that it must be shared with both people and nature, and that long-term sustainability must become the primary long-range goal.

Jim and I have had many conversations over the years about how we have done agroecology together. From our collaboration over 30 years, we have developed our strong belief in the need for whole food system change. We have learned together that agroecology is about the broad integration of research, farming practice, and social change actions. Without all three, it is not really agroecology.

Acknowledgements

This case study is adapted from material presented in Chapter 22 of the third edition of my textbook, *Agroecology: The Ecology of Sustainable Food Systems.* I especially appreciate the open exchange and relationship that I have with Jim Cochran and his Swanton Berry Farm operation. And without the contribution of a large number of collaborating researchers and students at UC Santa Cruz, this participatory relationship would not have been possible.

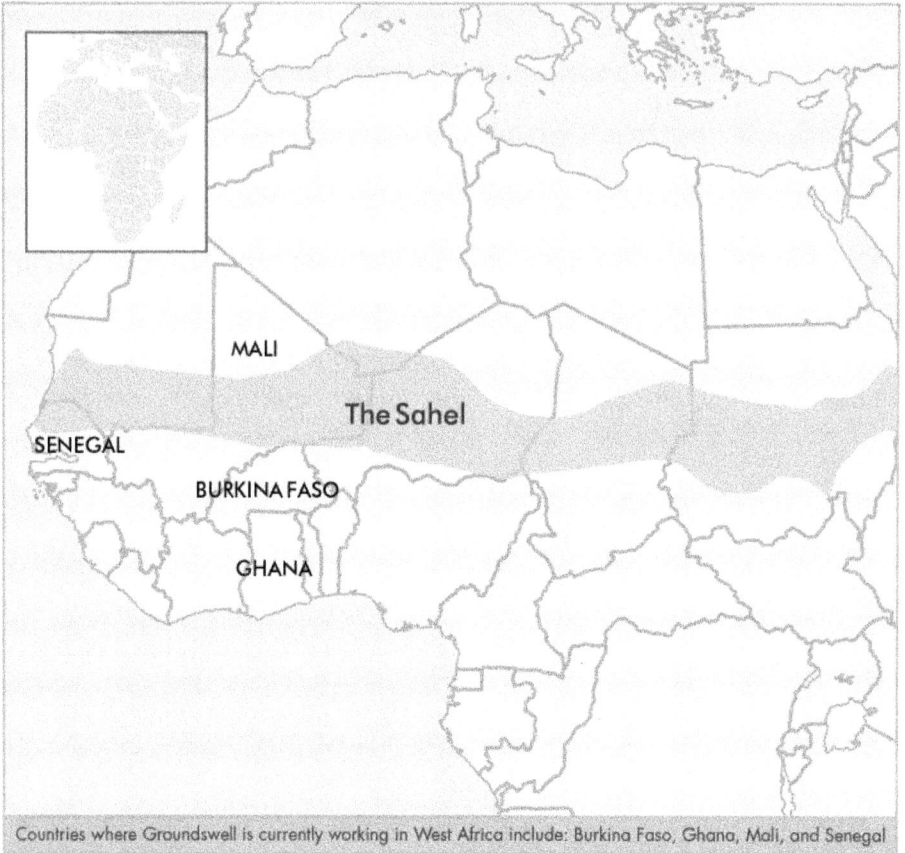

MALI

The Sahel

SENEGAL

BURKINA FASO

GHANA

Countries where Groundswell is currently working in West Africa include: Burkina Faso, Ghana, Mali, and Senegal

CHAPTER 6

West Africa context: Challenges facing family farmers in the Sahel

Peter Gubbels and Steve Brescia

Summary: *Rural families in West Africa's Sahel are facing a common set of challenges. A perfect storm of demographic, economic, and ecological pressures have pushed them to abandon traditional tree and shrub fallowing practices that previously supported sustainable food production. Now, their soils are depleted, and many people face chronic hunger and food insecurity. As a response, West African governments have pledged 10 percent of their national budgets to support agriculture, but the question remains whether this promise will support the same cycle of soil depletion and hunger, or if it can be leveraged to support an agroecological transition. This context frames the work of the three processes of agroecological innovation described in the following case studies on Burkina Faso, Mali, and Ghana.*

Hunger, fallowing, and soil fertility

Every year, since 2012, nearly a quarter of the people in the Sahel—more than 20 million people out of a total of 86.8 million (Eijkennar, 2015; USAID, 2014a; Haub and Kaneda, 2014)—have faced serious to extreme hunger and under-nutrition. This is a dramatic increase from previous patterns dating back to 1966 (see Figure 6.1). The majority of the people facing hunger are small-scale farmers who depend on agriculture by growing dry-land crops such as millet, sorghum, and cowpeas (IRIN, 2008; Mathys et al, n.d.).[1] More than 20 percent of farm households in the Sahel are now living on less than 0.50 USD a day—the definition of "ultra poor" (Eijkennar, 2015).

In the past, rural families had strategies to maintain soils and survive periodic droughts. These strategies are no longer working and rural communities are caught in a downward spiral of low crop yields, hunger, and malnutrition. A failing agricultural development paradigm is contributing to a loss of resilience of farming communities—their abilities, strategies, and assets to respond to temporary crises are eroded. The underlying cause of this growing crisis of food insecurity is declining soil fertility.

For centuries, small-scale farmers in the Sahel and savannah zones in West Africa maintained soil fertility by employing a strategy of natural fallowing. After four or five years of cultivating a field, a farm family would clear new

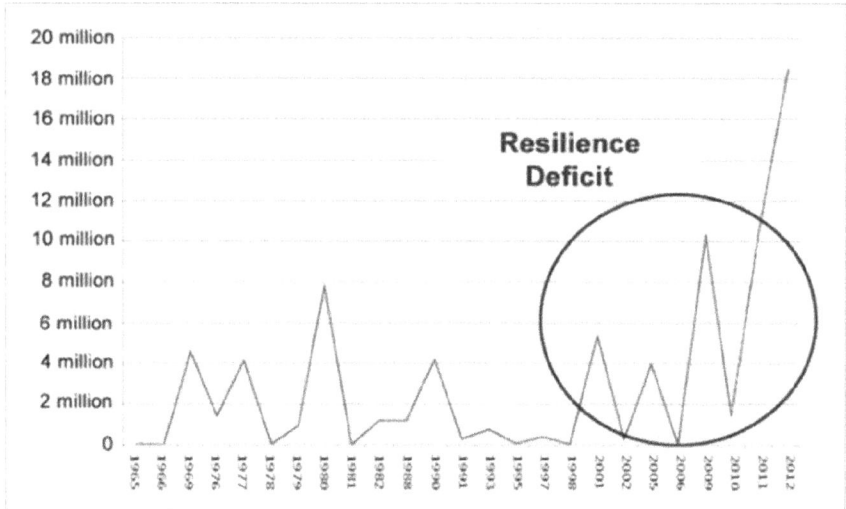

Figure 6.1 Number of people affected by drought and hunger in six sahel countries (1965–2011).
Source: USAID, Sahel JPC strategic plan: Reducing Risk, building resilience and facilitating inclusive economic growth, 2012, 2.

land, and leave the original field to fallow for 10 or more years. Even though farmers cleared land for planting by cutting and burning off trees and shrubs, these would regenerate from the living web of roots and stumps lying beneath the surface when the area was left to fallow. Given enough years, the natural re-vegetation of trees and shrubs would slowly restore soil organic matter and fertility by bringing up nutrients from lower soil layers, providing shade, and producing leaf litter. Eventually, families could cycle back to cultivate the land again.

In this way, peasant families worked with nature, managing the regenerative dynamic of trees and shrubs to sustain both their natural resource base and livelihoods. Today however, this practice of natural tree and shrub fallowing has all but disappeared. Since 1970, the population in the Sahel and savannah zones of West Africa has more than doubled. Land holdings have, consequently, shrunk in size. Farmers have been pushed to reduce fallow periods because they have less available land to cycle through. As they cultivate the same land year after year, they are removing more nutrients from the soil than are being returned. The Food and Agriculture Organization (FAO) estimates that 80 percent of the Sahel's land is now depleted of vital nutrients (Steyn, 2015; IRIN, 2008).

These exhausted, infertile soils result in lower crop yields, meaning that each family must cultivate a greater area of land just to produce the same amount of food. To do so, they have cleared more land and are greatly expanding the area under cultivation, further contributing to a vicious cycle of reduced vegetative cover. In some cases, the increased use of ploughs (pulled either by animals, or, in northern Ghana, tractors) has uprooted and greatly diminished the underground stock of living tree stumps and roots, thus further degrading the regenerative potential of the land. Farmers also continue to fell trees to provide fodder for their livestock, and to meet the timber and fuel wood needs of rural and urban populations.

An additional pressure in many areas of the Sahel is that the land is also shared with nomadic groups who herd cattle, sheep, and goats. These pastoralists face the same constraints of reduced access to land and vegetation to maintain their herds. Their need for the same dwindling resources has led to rising tensions with the settled farmers.

As the older trees in farmers' fields die off, new trees are not replacing them. Studies by the International Centre for Agroforestry (ICRAF) verify that since the 1970s and 1980s, farmers in the Sahel have experienced massive tree losses from drought and human population pressures. There is just not enough organic matter to maintain soils and feed people and livestock.

Fragile soils are exposed to wind and water erosion when battered by the torrential storms of the short four-month rainy season. In some cases in the drylands, the topsoil has been almost completely stripped away. A thick, almost cement-like crust develops, making it harder for rainwater to soak in and for germinating crops to emerge.

Finally, climate change is making rainfall patterns increasingly erratic. At times there is not enough rain, at others too much, or it often falls at the wrong time, delaying or shortening the growing season and leading to crop failure (IPCC, 2008). Most alarming, scientists project a temperature rise of three to five degrees Celsius above today's already high temperatures by 2050. Crop output could plummet if temperatures rise above a tipping point. With maize, for example, there is a 0.7 percent decline in crop production for each 24 hours exposure to a temperature above 29 degrees Celsius (84.2 degrees Fahrenheit). By 2050, scientists predict a decrease in agricultural production of 13 percent in Burkina Faso, 25.9 percent in Mali and 44.7 percent in Senegal (Potts et al, 2013). Even if overall rainfall remains the same, decreased soil moisture caused by the increased evaporation connected to higher temperatures will threaten crop yields.

Mobilizing responses

In response to the growing food security crisis in the drylands, the international community needs to raise nearly US$2 billion a year for humanitarian assistance for just nine countries from 2014–2016. Governments in the Sahel have pledged to increase support for agriculture to 10 percent of their national budgets. Major international donors and aid agencies, such as the Gates Foundation, the Alliance for a Green Revolution in Africa (AGRA), the World Bank, the US government's Feed the Future Program, and the New Alliance for Food Security and Nutrition, which includes governments, multinational corporate agricultural input suppliers, and civil society representation, have made major commitments to develop agriculture and lift tens of millions out of poverty.

But the critical issue is: what type of agriculture should be supported that is best suited for the drylands? What pathways, programs, and policies will enable small-scale family farmers to sustainably increase productivity, lift themselves out of hunger and poverty, and become resilient to climate change?

Unfortunately, most African governments and international donors are predominantly supporting the modernization of agriculture through a "new green revolution." This model is primarily based on chemical fertilizers, pesticides, and herbicides; hybrid and genetically modified seeds (GMOs); and mechanization and irrigation. While some are providing limited support to build resilience in the worst affected areas through more sustainable strategies, in general they are investing little in agroecological approaches or research to determine their relative effectiveness. The overall goal of these programs is to address the hunger crisis faced by smallholder farmers, yet in practice they tend to invest where good returns on investments can be generated. These conventional agriculture strategies are often applied on the best land, with medium and larger scale farmers who have more resources, to support the production of export crops such as cotton, irrigated rice, or peanuts. The assumption seems to be that overall levels of increased national production will generate benefits that will trickle down to the poorest.

The reality is that these approaches are not effective for the mass of smallholder farmers living in ecologically fragile, risk-prone dryland areas, who earn less than US$1 or US$2 per day. If the goal is to alleviate hunger and poverty, then efforts must focus on these people. They cannot afford expensive industrial agro-inputs, and even those smallholders who do gain access to them through subsidies are cultivating soil that is so degraded that using fertilizer is only marginally profitable—at best. Even for somewhat larger scale farmers, the price of chemical fertilizers makes it economically irrational to use them for staple food crops. Rather, they use them for high value export crops, but even then they are vulnerable to both climate and market volatility. Investing in external inputs puts them at higher risk of falling into debt if they lose their crops or markets. Most fundamentally, chemical fertilizers do not address the underlying problem of generating more organic matter and improving the biological health and innate fertility of soils.

Many major international development agencies seek to expand conventional agriculture and actively work to shape the perspectives, programs, and policies of national governments and agriculture ministries. Those responsible for implementing agricultural development programs lack knowledge of agroecological farming and its benefits, but also of the negative impacts of conventional and "new green revolution" agriculture. A common belief and bias of policy makers and donors, for example, is that integrating trees into fields (agroforestry) will limit productivity because they inhibit the use of machinery for cultivating "modernized" mono-crops. They believe that chemical fertilizers are required to boost production. Such perspectives influence many farmers as well.

Another major problem is that agricultural extensionists promote these industrial agricultural techniques through top-down and "one size fits all" approaches to technology transfer. The complex and diverse needs of farmers and communities are ignored in favor of supplying pre-defined technical packages (Watt, 2012). Traditional extension work generally does not adequately take into account farmers' norms, attitudes, objectives, and differing resource levels (USAID, 2014b). Rapid "scaling up" assumes homogenous farming populations and conditions, which does not reflect the local realities.

These biases are captured in an analysis of agriculture in the Sahel by the Institute for Development Studies, which applied a political economy perspective. They identified a triple neglect: of *agriculture* as a sector; of the needs of *small-holders* in marginal areas ill-served by green revolution technologies; and of an *alternative, multi-functional approach* to agriculture better adapted to millions of dry land farmers. This neglect persists, according to the report, because of the lack of political will, the lack of capacity of smallholder farmers to exert a strong demand for appropriate agricultural services, and ineffective strategies to address the complexities involved in scaling up agroecological innovations (Watt, 2012).

To counter this, Dennis Garrity, the Chair of the EverGreen Agriculture Partnership and UN Drylands Ambassador, has emphasized the need to embed the concepts of agroecological farming into the hearts and minds of conventional agronomists and policy makers who still see larger scale industrial agriculture as the solution to food security in West Africa. In particular, Dr. Garrity called for efforts to dramatically scale up the use of trees in cropping and grazing lands of smallholder farmers.

There is growing evidence that an ecological intensification of farming, building on traditional farming practices and principles such as fallowing, is the way forward. One proven and promising approach is *farmer-managed natural regeneration of trees* (FMNR), in which farmers integrate trees into their farming systems. In effect, this agroecological strategy amounts to "simultaneous fallowing," or taking advantage of the dynamics of fallowing while cultivating food crops.

Agroecological farming principles are essential in enabling smallholder farmers in West Africa to overcome the linked crises of soil fertility collapse and hunger that they are facing. Applying these principles at a larger scale, however, remains a challenge. The conventional "technology transfer" strategy influences and biases even those organizations seeking to scale-out agroecological practices.

This is the general context for the three following case studies from Burkina Faso, Mali, and northern Ghana. In each one, local NGOs, farmers organizations, and other scientific, government, and civil society allies have undertaken initiatives to achieve widespread adaptation and adoption of

Photo 6.1 Woman farmer standing in her FMNR managed plot in eastern Burkina Faso. *Photo credit:* Tsuamba Bourgou

improved practices required to enable small scale farmers to progressively make a transition to more agroecological, productive, and resilient farming. They have also worked to create more supportive policies.

Note

1.   Estimates of the percentage of the population who are small scale farmers varies by country, but most indicate this group constitutes at least 50% to 60% of the total.

Regions referenced in chapter

CHAPTER 7

Regenerating trees, landscapes and livelihoods in Mali: A case of farmer-managed transformation

Pierre Dembélé, Drissa Gana, Peter Gubbels, and Steve Brescia

Summary: *During the 1960s-1990s in Mali, centralized, national environmental management and population pressure ushered in massive soil degradation and the spread of national food insecurity. In the Mopti region, communities mobilized to stop the cycle of desertification and vulnerability by restoring traditional farming practices and village organization. Their work, in combination with decentralized government control and NGO support, has allowed them to spread agroecological strategies in Mali and beyond. Here, from the point of view of their NGO allies, we tell their story.*

History: From local environmental management to centralized control

Salif Aly Guindo understands the powerful relationship between trees and community organization in the challenging ecosystem where he lives. Just to the north of his community of Ende in central Mali, stretches the vast Sahara desert. "Our rainy season for farming here is less than three months," he reflected as he stood among one of the many thickets of trees that now dot the farmland of Ende. "We produce with our own seeds, conserved by our ancestors. Even if we are short of food, we will buy food from outside, but we always keep our own seeds. But before the *Barahogon* became strong again, the soil had become so poor that our seeds were not producing adequately. Now that we have been able to restore the soil, our seeds are producing again."[1] Salif is the President of the *Barahogon*, a traditional community organization that is regenerating the tree cover and soil fertility, and, as a result, reversing and holding back the spread of desertification from the north that had threatened to make Ende unlivable.

Since settling in the area in the twelfth century, the Dogon, Fulani, and Dafing peoples have farmed millet, sorghum, and cowpeas and raised livestock. The area is now known as Bankass *Cercle* (an administrative subdivision) in the Mopti Region of Mali. The dryland ecosystem, where annual rainfall varies

Photo 7.1 Salif Aly Guindo, Barahogon leader, Ende, Mali.
Photo Credit: Steve Brescia

between 400 to 850 mm per year, is challenging, but through a combination of well-suited social structures and ecologically appropriate farming practices, these people built a successful society. By 2009, the population within the Bankass *Cercle* had grown to 263,446 inhabitants (République de Mali Institut National de la Statistique, 2010).

For centuries, local Dogon communities were organized under the leadership of a *hogon*, or king, who oversaw multiple administrative departments. One of these departments was called *Barahogon*, which in the Dogon language means "King of the Bush." The *Barahogon's* traditional mandate was to sustainably manage the balance between production, livelihoods, and natural resources. The individuals working in this unit developed a deep level of knowledge about their local ecosystem, integrating it into their cultural practices. They were responsible for monitoring the bush lands; conserving important species of plants, trees and animals; enforcing local regulations for harvesting wild fruits; and setting the calendar for traditional festivals, tree-pruning, and seasonal confinement and letting loose of village animals for grazing. The members of the *Barahogon* were also responsible for communicating and enforcing traditional laws for environmental regulation, included directives to not kill female, pregnant, or lactating animals; to not cut down fruit trees; to use good hygiene practices around water sources; and to resolve local conflicts over land and environmental resources. Like most other traditional institutions engaged in environmental management, the *Barahogon* was eroded during colonization and, increasingly, when Mali gained independence from France in 1960. Over the next three decades, the Malian government centralized environmental regulation and management, shifting control from local communities to the newly created Department of Water and Forests. While the Malian state's intent was to promote more sustainable environmental management, their measures actually had the opposite effect.

Accelerating degradation, rising hunger

Between the 1960s and 1990s, a variety of factors combined to create a disastrous cycle of accelerating soil degradation, loss of tree and vegetative cover, and less-predictable rainfall in in the Bankass *Cercle,* the Mopti Region, and the wider dryland areas of Mali. Key drivers of land degradation included increased population pressure; the centralization of political power and decision making over local management of natural resources; the limited capacity of community organizations to generate agroecological innovations rapidly enough to keep up with the accelerating pace of environmental decline; and climate change.

For the previous 900 years, the traditional farming and environmental management practices of the Dogon and other ethnic groups in Bankass had enabled them to sustainably meet their needs, while coping with occasional droughts and bad years. The process of community-based management of natural resources worked. One of their most important practices was field fallowing (described in more detail in Chapter 6—West Africa Context

Photo 7.2 Cleared land for planting.
Photo credit: Steve Brescia

Chapter). When soil fertility began to decline in a field, farmers would leave it to fallow for 10 to 20 years, and would clear new land to cultivate. During this fallowing period, vegetative and tree cover was restored, adding new organic matter to the soil and making it fertile once again, for future cultivation.

While the *Barahogon's* capacity and authority to manage and encourage sustainable practices was gradually decreasing, the population was gradually growing, putting increased pressure on the land. As a result, in many parts of Bankass, farmers had to continuously reduce their number of years of fallowing. Some stopped the practice altogether. Because the *Barahogon* power had declined, they could not step in to promote more sustainable and beneficial response to these pressures. As a result, soil fertility and crop yields declined, and farmers found themselves in a vicious cycle of reduced fallowing time, declining soil fertility and water retention, the need to compensate by cultivating more land, decreasing tree and vegetative cover, and increasing erosion. Trees, also a traditional source of cooking fuel, had become so scarce that women had to travel great distances to gather firewood. Unable to find firewood, Salif Aly Guindo recalls many resorted to using "cow dung and sorghum stalks to burn for cooking fuel. That is what used to fertilize the land. But if you burn it to cook [instead of reincorporating it into the soil], then nothing will grow." Soil degradation intensified.

As the Dogon farmers became much more vulnerable to food insecurity, climate change also started to increase the frequency and severity of droughts. The severe Sahelian droughts of 1973 and 1985 in particular, and five more between 1992 and 2005, as well as the locust invasion of 2004-2005, all further accelerated the destruction of the vegetative cover. Wind erosion increased, and for the first time sand dunes emerged in the Bankass area.

Facing crisis, local leaders took action. They worked to reinvigorate the *Barahogon* and their cultural practices of sustainable land management. They acted to change the laws that had become an obstacle to local agency, control, and innovation.

Agroforestry practices: The power of law

Forestry laws established by Mali's newly independent national government during the 1960s did not adequately recognize farmers rights and needs to manage trees on their farmland, or participate in sustainably managing the bush. The same laws intended to protect the bush, also required farmers to obtain permits from the Department of Water and Forests if they wanted to cut or prune trees on their own land. This took away farmers decision making, and created bureaucratic process that dis-incentivized farmers from actively maintaining trees on their farms.

The Malian government recognized the need for reform in the early 1990s, and took important legal and legislative steps to again decentralize powers to regional and community authorities. For example, the government revised the National Forest Code in 1995 to categorize trees under ten years of age located on farmers' fields or fallowed land as a part of "agriculture" rather than "forestry." While this reduced restrictions on farmers, the Code left important details unclear. Were farmers allowed to prune and cut trees? What was the exact division of responsibilities between local communities and national government ministries?

The importance of culture

Around the same time, in 1994, a UK-based environmental NGO called SOS Sahel began working with farmers in the Bankass area to support communities to address the crises they were facing by promoting soil and water conservation and related agroecological techniques. As Drissa Gana, one of the authors of this case study, reported to Groundswell: "We started working in the area by undertaking participatory diagnostic studies with communities on the environmental challenges in the area. We gradually became aware in the oldest Dogon villages of traditional institutions for environmental protection that had existed, including the *Barahogon* in the north of Bankass, and the *Alamodiou* in the south" (Gana, 2015).

Another SOS Sahel staff person, Mamadou Diakité, observed how communities were beginning to work to promote traditional agroforestry techniques, and the value certain trees had in local culture. The *balazan* tree, as it is locally known (*Faidherbia albida*, or *Acacia)*, is native to the region and has multiple beneficial characteristics. It has a deep taproot that allows it to resist and survive drought, is thorny to protect it from animals, and is leguminous. The *balazan* holds soils in place, draws up nutrients from underground to regenerate and maintain soil fertility, contributes organic matter to soils through leaf litter, and fixes nitrogen. The tree's great advantage to farmers for integrating into their fields is that it loses its leaves in the rainy season, when they are planting, so it does not compete with crops for sun or nutrients during the growing season. Not surprisingly, the *balazan* tree has a prominent place as a source of life in the creation stories of local cultures.

To celebrate and encourage the recovery of agroforestry with *balazan* trees, and tap into these deep cultural roots, Diakité wrote a poem about the mutual dependence between peasants and the *balazan* trees. The poem was widely disseminated on local radio stations, and resonated strongly throughout the local population (Diakité, 1995):

Call of the peasant to the Balanzan:[2]

Balanzan tree, (don't leave me) don't abandon me ... Balanzan, don't abandon me

Balanzan, you who protect the farms against wind and the great heat of the dry season.

Balanzan, don't abandon me ...

Peasant from Seno, come help me out, be my hope to save me;
Peasant from Seno, protect me against hoes

Peasant from Seno, protect me against the plows

Peasant from Seno, protect me against the axe cuts of the herders
Peasant from Seno, protect me against the bush fires

But after a few of years of supporting farmers in the area, SOS Sahel had to discontinue the work and focus elsewhere in Mali. Yet farmer leaders of the Dogon villages continued their work to restore the environment and livelihoods on their own. In 2004, SOS Sahel underwent structural changes in its UK headquarters, and supported the Mali-based staff to establish a local NGO to continue the important work, which is now known as Sahel Eco. In 2005, after an absence of about eight years, Sahel Eco staff once again visited the area to learn about what had been happening. They were amazed by the changes. "When our staff returned to Ende in 2005, in addition to the few mature trees that had survived the droughts of the 1970s and 80, we saw a forest of young trees emerging on the sandy soils. We realized that something significant was happening," said Drissa Gana (Gana, 2015).

The *Barahogon* revitalizes itself

Recognizing the crisis they were facing, several communities in the Mopti region had decided to take radical action in the mid-1990s to revitalize their traditional organizational structures and sustainable land management practices. They worked to strengthen the *Barahogon* association, which covered the three districts (*cercles*) of Kani Bozon, Bankass and Koporo Na, around the village of Ende. *Barahogan* leaders worked with families both to recover past cultural practices and knowledge that had been effective, as well as to innovate in the face of accelerating climate change.

The resuscitated *Barahogon*, led by community leaders like Salif Aly Guindo, instituted simple community guidelines stating that no one could cut trees or branches in a field without first receiving permission from the farmer. They made announcements on local radio to inform surrounding villages of these rules, and let woodcutters know that their permits were only valid on state-controlled forestlands. They promoted a number of changes in the practices of local people. As Pierre Dembélé, Director of Sahel Eco and co-author of this case recalls, "Sometimes the people resisted these changes, but community leaders realized they were necessary in order to regenerate the trees, protect the trees, and regenerate the soil" (Dembélé, 2015).

Farmers had to learn new strategies to regenerate trees on their land and adjust farming practices to integrate trees into their farming systems. They learned to identify and select young shoots from living roots and stumps, mark them, and allow them to grow. They experimented with properly pruning growing trees in order to maximize their benefits and control shade

Photo 7.3 Formerly barren fields in Ende, Mali restored through farmer managed natural regeneration (FMNR), thus increasing the production of food crops, fodder, and fuel wood. *Photo credit:* Steve Brescia

to allow adequate sunlight for crops. They tested what different combinations of tree varieties were best on their farms, the uses and benefits of different varieties (e.g., soil fertility, fodder, wood for fuel, tools or construction, etc.), and the best density levels of trees per hectare. Women who were responsible for gathering fuel wood and fodder had to accept that for the first one to three years they had to allow the trees grow until they could be used sustainably for those purposes. As Salif Aly Guindo remarked in 2011, "You see how these trees have grown. If we prune all the small branches at the rainy season, we will get a lot of wood. This helps women because they are able to carry it on donkey carts to their houses. They don't have to travel as far." The *Barahogon* also managed disagreements between farmers, woodcutters, and government officials. In doing so, they had to work to generate legitimacy and trust within the community to ensure that these new rules would be accepted by all.

   In a relatively short period of about ten years, the *Barahogon* association around Ende achieved an incredible result: major ecological restoration and transformation of their lands, through their own innovation, organization and work, following principles of farming and living in balance with their natural resources. They not only stopped, but actually *reversed* soil and environmental

Table 7.1 Changes in practices identified by farmers

Favorable Practices to Increase	Unfavorable Practices to Decrease
Saving young tree shoots when planting crops	Destroying tree seedlings
Improved strategies to clean fields	Cleaning and burning of branches and stumps
Production of tree seedlings	Gathering crop residue and stalks (instead of integrating into soil)
Selling tree seedlings	Commercial exploitation of roots and bark
Planting of trees	Exploiting roots for artisanry
Pruning trees	Mutilating trees
Gather tree cuttings (for fodder and fuel wood)	Logging firewood from farm and fallow fields
Using improved stoves	Using millet stalks for fuel/cooking
Improved fruit-picking methods	Picking immature fruit
Plowing perpendicular to the slope	Plowing parallel to the slope
Mulching techniques	
Semi-direct integration of manure	
Managing animals in the fields	

degradation within their landscape—over 3,000 hectares. Between 1999 and 2008, they led this work without any external support.

Understanding and scaling success: NGO-*Barahogon* partnership

Seeing how much progress and agroecological transformation the *Barahogon* had generated, Sahel Eco decided to help them consolidate this work and spread it more widely. They invested in studies to understand and document the processes and agroforestry techniques developed by the *Barahogon* in Ende. Then, they began to collaborate with other villages around Ende to support them to inspire, revitalize, and train members of the other, largely dormant traditional associations across the entire Bankass area. They used techniques such as learning exchange visits, local radio programs, contests in which the most successful farmers were awarded prizes, and strengthening of traditional, community based farmers' organizations. This propelled a scaling-out of agroecological innovations to a critical mass throughout the Bankass *Cercle*. The set of agroecological techniques developed independently by the *Barahogon* was eventually termed "Farmer Managed Natural Regeneration of Trees" (FMNR; or in French, *Régénération Naturelle Assistée*, or RNA), and has been promoted widely beyond *Bankass* and Mali (as demonstrated by the case studies on Burkina Faso and Ghana).

As one step in the scaling process, Sahel Eco worked intensively with the new decentralized local government authorities in Bankass, pushing for legal recognition of the *Barahogon* and *Alamodiou* inter-village associations.

Photo 7.4 Restoration of tree cover on once desertified landscape around the village of Ende, Mali. *Photo credit:* Steve Brescia

Then, Sahel Eco worked to foster official agreements *(conventions locales)* between local government authorities and the *Barahogon*, which defined the communities' responsibilities for managing natural resources.

Sahel Eco actively promoted the spread of these farmer-led agroecological practices throughout Mali, and soon, the village of Ende became the host for scores of group visits – from other villages, government officials, national and international NGOs, and delegations from other Sahelian countries—to see the remarkable work of the *Barahogon* association. Sahel Eco worked to link famer-to-farmer and community-to-community spread of FMNR and complementary sustainable practices to wider efforts to influence policy and strengthen national and regional networks.

Participatory research, documentation and communications

Sahel Eco worked with community members to carry out participatory research and analysis, and wrote and disseminated a number of articles and studies. Short videos as well as longer documentary films were produced and shared with other communities and civil society and farmers' networks to inform them on FMNR approaches and benefits. Sahel Eco and the *Barahogon* also used local radio to promote stories and poems,

---

Dialogue During 2009 Cross Visit to Ende:

"I would like to know how you have been able to water such a large plantation (of trees) when the closest water point is 4 km away?"
   Naomie Dembele, farmer from Tominian
   "Madam, the trees that you see have not grown because they have been watered. The growth of this forest is the result of two activities. First, we ensured protection of the area against devastating logging through an agreement with the forest services. And secondly, many farmers applied the technology of FMNR."
   Salif Aly Guindo President of Barahogon association.

---

share effective practices, and inform community members of regulations for managing trees.

Below are a few examples of participatory strategies carried out by Sahel Eco to generate knowledge and change practices:

Cross Visits: Sahel Eco promoted learning visits to FMNR work in Bankass by village delegates from other *cercles* in the Mopti region (Koro, Bandiagara, Douentza, and Mopti) and in the neighboring Tominian *Cercle* of the Segou Region. For example, in 2009 Sahel Eco organized a special "caravan" to visit farmers and local government decision makers from the Tominian *Cercle*, who traveled in several buses to Ende, Bankass. Four farmer organizations from Tominian selected 61 participants, including 25 women, who agreed to be responsible for sharing what they had learned, and teaching and training others upon their return. The caravan visit was a powerful catalyst in spreading the work to Tominian.

Competitions and Prizes: Sahel Eco and the *Barahogon* organized competitions for the "Best FMNR Farmers," promoted over radio and by word of mouth among community-based organizations. In 2010 for example, 228 farmers joined the competition, and 50 winners were selected from ten municipalities in the Mopti region. Each winner received a bolt of cloth, prized for making clothing, printed with "Greening the Sahel" designs. They made videos of seven winning farmers, which community promoters then used to train others.

Advocacy and "Local Conventions:" As noted, advocating for decentralized control and decision-making over the use of land and trees was a crucial factor enabling the successful spread of FMNR. Sahel Eco also helped to facilitate the creation of local agreements in all 12 towns in Bankass through participatory action-research and negotiation processes involving local stakeholders. These agreements formalized farmers' local control over managing trees. They included incentives for farmers to regenerate and maintain trees, confident they will be able to derive long-term benefits.

Networking: Sahel Eco, with the support of international partners and funders, developed programs to spread the strategies beyond Bankass to the *cercles* of Koro, Bandiagara, Mopti and Djenne. They created wider networks with researchers and representatives of NGOs and farmers' organizations that shared similar goals. Sahel Eco also discovered that parallel FMNR efforts were emerging in other countries in response to similar dynamics and crises. For example, researchers and organizations from Niger had identified about 5 million hectares of agricultural land regenerated by FMNR since the 1980s. Organizations from northern Burkina Faso had regenerated another 300,000 hectares.

In April of 2009, these groups gathered at a meeting to launch a wider "Greening the Sahel Initiative." In addition to Sahel Eco and farmers' organizations, the meeting included representatives of Mali's Ministry of Environment, the President of Mali's National Assembly, the National Coordination of Peasant Organizations (CNOP), and many NGOs, field level technicians, and representatives from other countries. Farmers gave convincing testimonies about their work to regenerate their land and livelihoods through FMNR.

Results

All of these strategies accumulated to create a self-spreading effect of FMNR among farmers and villages in many areas, as observed by external evaluators and local community members and officials. Such self-spreading and voluntary adoption are often the best indicators that appropriate technologies are bringing real and valued benefits to people's lives.

Overall, the work started by the *Barahogon*, and supported by Sahel Eco, has resulted in a remarkable reversal of deforestation and desertification, the spreading of FMNR to a critical mass of farmers in the Mopti Region, and significant contributions made to a wider Greening the Sahel movement in Mali and in West Africa. The results are clear and tangible to those living and farming in the Mopti. As Salif Aly Guindo describes:

> "Before there was no grass here. There was no fertility to the soil so it would not grow. Now, whatever you plant here will grow ...
>
> Before there were a just a few isolated, old trees. Now, when we plant sorghum, millet, cow peas, they all grow well. The falling leaves fertilize the soil. Those *nabana* trees (*Piliostigma reticulatum*) trees fix nitrogen, and so do those small trees there. Before 1999, the wind was a big problem. The blowing sand would cover the seeds and nothing would grow, and there were insect invasions. But now the density of trees stops the wind so the seeds can grow. We are also harvesting the grass for animal feed so we can raise more livestock.
>
> Before 1999, we harvested 4-5 sacks (100 kilos per sack) of millet or sorghum here. Now we get four to five granaries full. Each granary holds 15 sacks (*note: an increase from 400-500 kilos to 6,000 kilos of millet or sorghum per year*).

We started this work in 1999. Now we have enough food for the whole year. There is no hungry season. Before, we were really food insecure. We had to rely on relatives for food from different areas.

With a lot of collective effort we have transformed this area."

Beyond this powerful account, two evaluations conducted by Sahel Eco highlight the scale of the results thus far. The first was completed in 2010, and focused on the process in the Bankass area from 1999-2010. It found an increase of 30 to 50 percent in the rates of adoption for ten "best practices" the farmers had identified for managing tree resources through FMNR (Gubbels et al, 2011). The adoption rate for most of these practices has passed a "critical mass" (estimated at 40%), which indicates the likelihood of self-sustaining adoption and continued spread to other farmers. Farmers had regained a deep awareness of the value and importance of trees for their lives and their landscape. As a result, among 13 *communes* in Bankass, six have now reversed deforestation and reached a stage of "balance" between their farming production and their regeneration and management of trees (i.e., increased wood resources meet or exceed annual consumption levels).

But how do these quantitative indicators map onto real benefits of FMNR for families? To further assess this, Sahel Eco and the *Barahogon* carried out in-depth focus groups with women and men in the communities of Plateau and Seno in Bankass *Cercle*. We found that in Plateau, for example, women who had been using millet stalks and cow dung for cooking five to ten years earlier were now using only sustainably harvested firewood from their fields. Table 7.2 summarizes the perceived change in benefits from FMNR for the people in Plateau from 2005 to 2010 (on a scale of 1 = lowest, and 10 = highest).

In Seno, participants assessed the changes in the state of their natural resources between 2000 and 2010 (see Table 7.3). Using a 1-10 scale, they identified the year when different tree resources were at their greatest level over the last ten years (scored as 10), and then identifying how resources had evolved during that period (with 1 as the lowest level of available tree resources).

Farmers also identified significant benefits in terms of how they had strengthened their local organizations and created a more enabling context. For example, through the establishment of *Conventions Locals* in most communes, they significantly improved community management of forests, and eventually established a Union for Inter-community Forest Management around the Samori forest area. Community members, as well as local government officials, expressed great enthusiasm and a strong sense of local ownership for the work. They created more transparency in the sale of licenses for cutting and using trees, increased municipal revenues from the sale of licenses, and reinvested these resources in local development. Fewer people were migrating out of the area, the loss of livestock had been reduced, and local craftspeople had again recovered artisanal, income generating activities with greater access to natural resources.

Table 7.2 FMNR benefits from 2005-2010 as defined by Plateau communities (on a 1–10 scale)

Indicators that a community gets a large portion of its needs met by local trees	Situation in 2005	Situation in 2010
Natural medicines	2	8
Fruit from trees for nourishment	3	7
Abundance of products	1	8
Firewood	3	7
Construction wood	2	10
Improvement of soil fertility	1	9
Return of wild animals	1	10
Leaves for cooking in sauces	1	10
Production of animal fodder	3	8
Shade	1	10
Abundance of trees	2	9
Reducing wind erosion	1	10
Increasing income	1	10
Return of birds	1	10
Introduction of exotic tree species	NA	10
Strengthen social cohesion	NA	10
Average	1.5	9.1

Table 7.3 Seno evolution of natural resources

Type of Tree Resources	Year 2000	Year 2005	Year 2010
Sacred forests	10	3	1
Groves	3	5	10
Hedgerows	2	3	10
Trees in public places	4	5	10
Trees in fallows	3	5	10
Trees in farm fields	3	5	10

The second evaluation, completed in 2013, focused on a program from 2010-2013 to expand FMNR to the commune of Sokura, in the Mopti Region (Gubbels et al, 2013). Among 11 (out of 28 total) villages surveyed, farmers increased their adoption rates of FMNR from 18.8 percent of households in 2010, to 41.3 percent in 2013. Farmers regenerated over 56,500 trees (12 different native species) through FMNR on their farms and fallow lands. They planted a total of 25,241 additional "high value" trees (17 different varieties including fruit trees, baobab, moringa, etc.) on farms, with a survival rate of over 96 percent.

In Sokura, community members identified many similar, positive benefits from the widespread adoption of FMNR. Women reduced their workload and expenses, and increased their access to firewood. Farmers improved soil fertility, while reducing soil erosion and sand blown by wind and storms, and also saw improved germination rates of crops in their fields. By using the increased animal fodder available from leaves and branches, they increased their production of meat and milk products, as well as their income by fattening and selling animals more quickly. Farmers also had increased access to non-timber forest products, such as fruits and natural medicines for use and sale. Women earned more income, for example by selling *Faidherbia abida* fruits (women reported gathering one 100 kg bag per tree in FMNR fields, which they sell for the equivalent of around US$4.40/bag). Families had more access to wood for beams and poles for use in house construction. Participants reported that increased access to and use of all of these resources resulted in improvements in their health.

The evaluation found that the impacts were stronger in 15 villages in rural dryland areas, as compared to the 13 villages that were either in peri-urban areas (due to commercial pressures and the insecurity of land tenure), or in rice growing flood plains (which are a different ecosystem with different farming patterns). A lesson is that we need to better adapt FMNR strategies to peri-urban areas where different market opportunities may exist, as well as to different agroecological contexts.

A rough cost-benefit analysis of the Sokura program demonstrates an impressive level of effectiveness and efficiency. The program represented an investment of approximately US$165,000 over three years, or about US$55,000 per year. For this investment, in at least 15 of 28 villages, around 40 percent of farmers were adopting FMNR, substantially reversing deforestation and desertification, and sustainably increasing soil fertility, food production, income generation, and multiple related benefits for women and families.

Lessons and next steps

The collapse of soil fertility and the rise of chronic vulnerability and hunger among rural populations are among the greatest challenges facing Mali and the wider Sahel region of Africa. Farmers and their traditional organizations, such as the *Barahogon,* have demonstrated a remarkable capacity to draw upon traditional knowledge, continue to innovate, work in balance with nature, and regenerate their landscapes and livelihoods. Based on the *Barahogon's* initial powerful track record of success with farmer managed natural regeneration of trees, Sahel Eco has worked effectively to spread this strategy more widely to other traditional organizations of the Bankass area. Just as importantly, they have worked with multiple allies to advocate for appropriate regulations and agreements to decentralize decision-making over trees and land management to farmers and communities. It is clear that once farmers feel free to protect, care for, and benefit from trees in their fields, organizational and technological solutions can emerge. FMNR and related agroecological techniques have great potential to help re-green the Sahel and reverse the cycle of degradation in the region.

In spite of this important progress, communities in many areas of Mali continue to experience high rates of deforestation and increasing vulnerability. Even while Mali's government has many policies that are favorable to agroecology, in practice they have not prioritized it's spread. Instead, government agricultural investments favor large-scale commercial agriculture and international agribusiness interests. For example, they provide access to highly productive land to foreign countries, and neglect the majority of small-scale farmers in the dryland areas.

To turn the tide and scale agroecology in Mali, a number of steps are required. Laws and regulations must continue to be revised to ensure that farmers are empowered to sustainably manage their trees and land, and to recognize traditional land ownership arrangements. Government agriculture and forestry departments, local authorities, NGOs, and farmers' and traditional organizations must come together and harmonize principles around these goals. Traditional organizations must continue to strengthen their capacity to manage their natural resources, while municipalities should create territorial development plans that delegate responsibilities to them. There is a need to support farmer-led innovation processes to integrate a wider array of agroecology principles and practices into existing FMNR practices to further increase production and resilience. Farmers and their allies must improve agroforestry product value chains for local markets, while research institutions should collaborate to support the development and documentation of all of these strategies.

In May of 2014, many organizations in Mali came together to create a national agroecology platform to pursue these goals more coherently. Learning from the inspiring successes such as that of the *Barahogon* in Ende, the lesson is clear: it is possible to transform farming systems that are in crisis to become productive, resilient, and sustainable. However, doing so requires allies in the agroecology movement in Mali to work together to ensure scaling from the ground up, while also creating enabling policies. As Salif puts it, "This requires a collective effort. Even with the creation of the *Barahogon*, you can see that some people still may try to come and cut some trees. We have to be organized. An individual farmer cannot do it alone."

Notes

1. Salif Aly Guindo. Interview by Steve Brescia, Fatoumata Batta, and Peter Gubbels, June 14, 2011
2. Balanzan is a local spelling of the "Balazan" tree.

MALI

NIGER

BURKINA FASO

Djibo

Ouahigouya

OUAGADOUGOU

Fada-Ngourmo

Tenkodogo

Bobo-Dioulasso

Banfora

GHANA

TOGO

BENIN

COTE D'IVOIRE

Regions referenced in chapter

CHAPTER 8

From oases to landscapes of success: Accelerating agroecological innovation in Burkina Faso

Fatoumata Batta and Tsuamba Bourgou

Summary: *The Association Nourrir Sans Détruire (ANSD, the Nourish Without Destroying Association) has been working with 125 villages in the Eastern Region of Burkina Faso to support a community-based, farmer-driven process of agroecological innovation and dissemination. Through field schools, exchanges, village-level action plans, and collaboration with many local leaders and government agencies, farmers and project collaborators have not only supported ongoing agroecological innovation, but found ways to spread innovation to an increasing number of farmers.*

Like many of his neighbors, Souobou Tiguidanla works in precarious environmental conditions to sustain a large extended family on a small farm where he primarily grows maize, millet, and sorghum. "In 2010 and 2011," Souobou recalls, "we were hungry. Rainfall was poor and we were not able to produce enough food for ourselves." The family could not survive on their own stores, and had to buy from the market, knowing that this expenditure would reduce their ability to invest in next year's crops. "Something needed to change," Souobou knew, so he sought out training in agroecology from a local organization, and gradually began to adapt new methods that were not only more productive and environmentally beneficial, but also less costly.

The *Association Nourrir Sans Détruire* (ANSD, the Nourish Without Destroying Association)—an organization which one of us (Bourgou) directs and the other (Batta) co-founded and partners with—has worked to support agroecological adaptation and innovation in the Eastern Region of Burkina Faso. By encouraging experimentation, recognizing innovation, and prioritizing decentralized farmer-to-farmer learning, ANSD has found pathways to more ecologically and economically viable livelihoods. We work toward widespread, lasting change by focusing on the **depth** of on-farm agroecological practice, the **horizontal** spread of practices from farmer-to-farmer, and the **vertical** adaptation of agroecology through layers of government and civic organizations.

Challenge: Oases of success

According to the 2015 UN Human Development Report, Burkina Faso is the sixth poorest country in the world (Jahan, 2015). The Eastern Region of the country is one of its most economically marginalized areas, and recent studies have estimated that 43.9 percent of the population lives below the poverty line (IMF, 2012). These people are caught in a vicious cycle of degrading natural resources, declining soil fertility, decreasing food production, and hunger. Food shortages are frequent, particularly during the "lean season" between harvests, and—as in Souobou's case—are made worse by drought. To survive, many families skip meals. The poorest 30 percent of smallholder farmers often sell their animals and other household assets during these periods in order to buy food from local markets. When they don't have anything to sell, they obtain high-interest loans from money-lenders. This asset stripping leaves households even more vulnerable for the next lean season or next drought. Most of the rural population of eastern Burkina Faso, as is also true in other parts of the Sahel, is unable to farm their way out of the this vicious cycle by relying on practices (such as fallowing) that had worked in the past.

In this challenging context, farmers, local NGOs, and agricultural researchers in Burkina Faso have developed a variety of viable solutions. Over the last 30 years, they have tested and adapted a number of effective agroecological farming practices—some new, others traditional—that have proven capable of restoring soil fertility and increasing food production for smallholder farmers. These include soil and water conservation techniques that build on traditional practices, such as "*zai,*" and "half-moon" micro-water catchment planting pits, and permeable rock contour barriers; the use of compost to increase organic matter in soils; and the promotion of "farmer managed natural regeneration" (FMNR) of trees. FMNR is an agroforestry approach in which, instead of clearing trees, farmers allow them to regenerate on their farms from existing stumps and roots, pruning the shoots and integrating the trees into their farming systems in a way that restores soil fertility and productivity. Some farm families have also adopted the use of short-cycle seeds to cope with irregular rainfall.

Although very effective, these approaches are currently only adopted on a limited basis. The farmers who do use them represent "oases of success" in a wider landscape of struggle. A much more dramatic and rapid spread

---

⬭ **Farmer testimony**

Adjima Thiombiano, Gayéri Village[1]

"The challenges we face are that the rain is insufficient and the soil is declining. Since the soil fertility has declined, the production has also declined. We don't have as many crops as in the past. There are 11 people in my household. Of course we are worried. If you're responsible for others and you don't have enough to eat, you're very worried."

Photo 8.1 ANSD animator demonstrating how to use the A-frame to create contour barriers for soil and water conservation.
Photo Credit: Tsuamba Bourgou

Isolated innovation

Mariam Ouango, a 57-year-old farmer and mother of six from the village of Tibga, has found an unusual way to increase crop production without using chemical fertilizers. In addition to farming and keeping livestock, Mariam also processes value-added products, such as shea butter. For many years, she struggled with raising tomatoes, which often ended up "burned" from limited rainfall and the harsh impacts of chemical fertilizers. One day, Mariam happened to notice that the land where she poured out the remains from her shea butter extraction process appeared to be less compact, more moist, and undamaged by termites and other insects. She sensed an opportunity, and began to experiment with spreading shea butter liquid in her garden in place of chemical fertilizers. The resulting tomato plants were twice as tall and productive. The new technique has had wonderful results for Mariam and her family. She is proud of her innovation, and motivated to continue experimenting with other agroecological strategies. So far, her innovation has yet to spread to other farmers. One limitation is that most farmers do not have access to shea butter extract. To determine if agricultural programs should focus on improving access and disseminating this technique, it will be important to carry out further research to verify the impact of using shea butter as a soil amendment.

---

of agroecology is essential to reverse the alarming degradation of soils and natural resources, regenerate productivity, and reduce poverty, vulnerability, and chronic hunger for peasant communities while creating greater well-being.

Response: Scaling-up to landscapes of success

ANSD's broad mission is to strengthen rural communities to overcome hunger and promote socio-economic development. In 2010, we initiated a new program in the Eastern Region to build on past work. This program promotes community-based agricultural development through agroecology in three districts that together have a total of 125 villages: Bilanga, Gayéri, and Tibga.

ANSD believes that farmers and their community-based organizations must lead their own agricultural and community development processes, and that our role is to sustainably strengthen their capacity. We also believe that individual projects must be linked to wider social change initiatives and policy-making. To do so, ANSD works in close collaboration with community-based farmers' organizations; two local NGOs, the Association for Research and Training in Agroecology (ARFA) and the Association for Rural Promotion Gulmu (APRG); the National Institute of Environment and Agricultural Research (INERA); and local government officials and traditional leaders.

As Fatoumata Batta (ANSD co-founder and co-author of this chapter) described in a report to Groundswell International:

> "When we started working in 2010, we realized there were farmer innovations that showed that agroecological practices were effective under even extreme conditions like those of eastern Burkina Faso, but they just were not spreading quickly enough to make a difference. We knew we had to find a way to 'scale out' agroecology. So we went to the villages and facilitated a number of discussions with farmers to understand why things moved so slowly. Villagers understood the problem very clearly. They said that while some had heard of these agroecological innovations, most farmers hadn't seen them or did not know a lot about them. There were almost no extension services supporting peasant farmers to learn about these alternatives. In general the government focuses on larger-scale farmers producing export crops, providing conventional inputs, and doesn't focus on small-scale farmers or sustainable approaches. Farmers analyzed that from their side, their communities generally lacked the organizational capacity to spearhead the promotion and spread of these strategies themselves, or advocate for them. High levels of illiteracy also made this a challenge.

> "We decided that we would work to support farmer experimentation with promising agroecological practices, and farmer-to-farmer spread of those practices. In addition to technical skills, this would also mean strengthening the organizational skills of the village organizations to lead the process. We made a commitment to prioritize the involvement of women leaders and women's groups with targeted strategies that made it easier for them to participate and benefit. We also planned to systematically strengthen the capacity of community organizations to create networks for sharing knowledge and effective practices across

many villages, to better access local markets, and to contribute to policies supporting food sovereignty" (Batta, 2015).

Villagers strategized with ANSD, developed activities and regular reflection sessions to assess and review their progress, identified lessons learned, and adjusted their strategies going forward. Starting small, ANSD used a multiplier strategy to spread agroecological practices, farmer-to-farmer and village-to-village, working in three complementary directions: first, we worked to **deepen** individual farmers' knowledge with a growing set of agroeco-logical principles and practices that diversified farmer innovation and experimentation. Then, project partners worked to spread this knowledge **horizontally** through farmer-to-farmer sharing and workshops. Finally, our team worked to **vertically** spread agroecology through wider farmer networks and policy work.

Agroecological depth: Drilling down with farmer knowledge

Farmers tend to listen to other farmers living in the same conditions—especially if they see things that are working. Therefore, ANSD organized learning visits for village farmer organization leaders, local government and ministry officials, and religious and traditional leaders to see successful agroeco-logical techniques practiced by innovative farmers.

Leaders worked with their villages to identify key challenges and oppor-tunities, and to determine ways to test and spread priority agroecological

Photo 8.2 Women in Bilanga-Yaga creating zai planting pits and adding compost.
Photo Credit: Amy Montalvo

innovations they had observed. Groups of farmers interested in trying the new agroecological techniques were formed. ANSD facilitated participatory organizational self-assessments with farmers' groups, and followed-up with appropriate support. This led to community members gradually establishing village agricultural committees in all of the 60 villages. One criterion of the committees was to ensure they had diverse representation in relation to gender, economic status, age, and traditional as well as religious organizations. These committees built their organizational capacity to take on the analysis, planning, awareness-raising, farmer-to-farmer coordination, and the monitoring and assessment of the agroecological sharing process.

Through this process, we chose "foundational" agroecology innovations: *zai* planting pits, stone contour bunds, half-moon water catchment areas, and FMNR/agroforestry to form the basis for the farmer-to-farmer technical and practical training given to the growing groups of farmers interested in agroecology.

As ANSD Director Tsuamba Bourgou described elsewhere of the process to spread and experiment with these techniques:

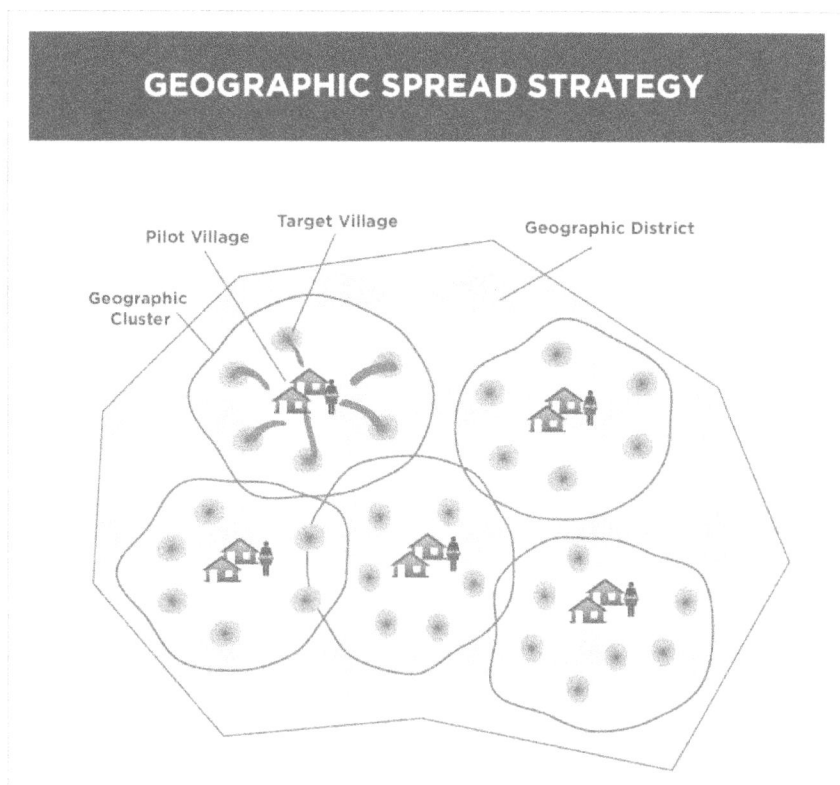

GEOGRAPHIC SPREAD STRATEGY

Figure 8.1 Geographic spread strategy for agroecology.

"We supported innovative farmers in pilot villages to conduct on-farm trials of agroecological techniques on test plots, assess the costs and benefits, and compare these to practices on the rest of their land. Then we organized the 60 involved villages into 17 clusters of three to four villages each, according to geographic proximity, socio-cultural group, clan affiliation, and use of the same markets. Community leaders designated a pilot village in each cluster, and within that village selected motivated farmers (women and men) to begin on-farm experiments in each major section of the village. These farmers used a farmer field school approach that experimented with a limited number of agroecological practices on test plots on their own land. We tracked the adoption of new practices with a simple, participatory system of community-managed monitoring and evaluation established in planning meetings that used visual tools like a social map of all households and easy to read charts that assessed levels of adoption" (Bourgou, 2015).

If not well designed, development programs can actually widen the gap between better-off and poorer households, or between men and women. ANSD worked with community members to address and avoid this by identifying the most vulnerable and food insecure households, prioritizing the participation of women and adapting targeted support to both. Community organizations

Photo 8.3 Farmer-to-farmer cross visit to learn agroecological techniques such as half moons.
Photo Credit: Tsuamba Bourgou

ensured that at least 30 percent of all farmers involved in key activities were women.

Spreading agroecology to more households

Agroecological practices spread beyond the participants of initial trainings through farm visits and experiments. To accelerate this spread, our teams created a plan to combine geographically dispersed pilot villages with farmer-to-farmer training. The village agricultural committees recruited an extensive, decentralized network of volunteer farmer leaders (both women and men) to teach others. These volunteers were selected based on their own interest and practice of agroecology, a desire to teach others, and geographic distribution to cover all communities. These volunteers are all part of a decentralized "cascading" approach to farmer-to-farmer training, organizing experimental plots in various villages to test the agroecological practices. When these farmer- volunteers were convinced of the efficacy of agroecological practices on their own farms, the village agricultural committees organized field days so that other farmers within village clusters could visit and learn from these experiments. Through ANSD, we provided methodological training so farmer-volunteers could effectively share their new practices and provide advice to a growing circle of interested farmers.

Additionally, farmer leaders worked with ANSD to develop community radio programs to share the benefits of specific agroecological techniques through local language broadcasts. They produced videos documenting local farmers' experiences to share them with other farmers and villages. Bourgou reported:

> "This is all part of our effort to work with farmers to generate, document, and disseminate knowledge in a vibrant and interactive way. At ANSD we complement community-led monitoring and evaluation with additional evaluation processes for program learning and to evidence generation. This information—and the information from our own, internal evaluations—contributes to the reports, videos, photos, case studies, and human-interest stories that we disseminate locally and internationally for use in promoting agroecology" (Bourgou, 2015).

Vertical Spread: Creating an enabling political and social context for agroecology

Although the government of Burkina Faso does provide limited support for some agroecological farming practices, it does not consider agroecology a development priority. Most public investments in agriculture are in the more high-yielding farming areas (such as cotton production), and they promote conventional practices delivered through technology packages of commercial seeds and subsidized chemical fertilizers. It is challenging to engage with and change national level policies—especially for rural citizens in ecologically fragile, risk prone areas who have limited influence. So, program participants first influenced the plans, budgets, and development priorities of local and regional governments and ministries. Most people in these agencies have a limited knowledge of agroecology. Involving them in the learning processes in rural communities helps develop a shared understanding and appreciation of agroecology and the farmer-to-farmer approach. Many become allies.

In 2014 ANSD convened two district workshops and one regional workshop where farmers and ANSD representatives shared lessons on agroecological strategies with farming organizations, local government officials, and decision makers to increase their awareness and discuss plans to strengthen and spread agroecology.

These meetings helped the ANSD and farmer leaders to systematically engage key actors at district and regional levels, allowing multiple agencies and organizations to share strategies and create a harmonized use of concepts and approaches to agroecology. It also provided more institutional support for the farmer-to-farmer visits, workshops, field days, and inter-village evaluation and planning sessions. Additionally, ANSD is an active member of several regional, national, and global networks, learning platforms, and communities of practice that support the spread of agroecology in Burkina Faso and beyond. All of these networks and learning activities help to reinforce agroecology within government agencies and open the door to more resources and favorable policies for agroecology.

Results: Increased agroecological innovation and adoption

After Souobou (the farmer described at the beginning of this chapter) attended an agroecological training event, he began to experiment with new practices on his farm. "I built stone contours on my fields," he said. "This keeps the rainwater from flowing away. We also started to make compost with crop residues and cow manure." As a result, his soil is more moist and fertile, and his yields have increased by over 100 percent in just one year. Souobou was the first farmer to adopt these particular agroecological practices in his village, but he will not be the only one for long. "My children are already learning to use the new practices and I am ready to teach others too," he said.

ANSD began its work in 2010 in only ten villages, but through the interest and efforts of farmers like Souobou, it has expanded to 60 villages. While the project began as a collaboration with community leaders and farmers, it now also works alongside the government agricultural research agency, INERA, to promote agricultural experimentation. Over half of the farmers involved in the program are now practicing agroforestry (FMNR), which was previously scarce in the region. Yearly crop rotation is also expanding.

Much of this success has been made possible by offering extensive opportunities in decentralized farmer experimentation and horizontal knowledge transfer. This creates commitment and ongoing interest in agroecology at a much lower cost than conventional agricultural development programs. The project has reached many people: between 2014 and 2015, a total of 221 farmer field schools were organized with over 2,500 farmer-trainers trained, most of them women. Thanks to the work of these farmer-trainers, a total of 2,945 households had adopted agroforestry (FMNR) and related agroecological techniques (*zai* planting pits, contour rock bunds, organic manure, etc.) by mid-2015. Other villagers then visited and learned from these farmers through organized field days. These events trained over 1,000 men and women across the 60 villages who act as *volunteer promoters*, providing agroecological training and follow-up support to farmers who are new to agroecology. The cost of providing training to these volunteer farmer promoters has been about US$2 per person. (Other NGOs and technical support organizations often spend more than US$10 per farmer trained in similar conditions, with less sustainable results.)

Through this farmer-to-farmer process, the program has been highly successful in creating a diverse base of agroecological farmers, leaders, and thinkers. After starting with the original farmer experimentation and farmer field schools, from 2010 to 2014 a total of 16,325 farmers, including 8,498 women, have participated in learning activities to gain a much deeper understanding of agroecological strategies, and are starting to adopt key agroecological practices. Future assessments will look at their levels of adoption and the impact on soil fertility and food production. Based on various program documents, we estimate that at least 3,000 children and youth have become engaged in environmental protection activities and many women have organized groups for agroecological practice and support. Together, all these farmers are part of a local collaborative movement that has been strengthened to spread agroecology.

One goal of the program has been to develop a "critical mass" of farmers in each village (30–40%) who are adopting agroecological techniques. When this occurs, we expect other farmers to begin to adopt practices organically without formal extension efforts. A 2014 survey (presented in Table 8.1) of 15 villages where the program had been working for four years revealed that this point has already been surpassed in regards to rotation (52.9%), FMNR (52.4%) and rock bund on the contour (40.1%).

We also seek to understand what agroecological practices farmers chose to adopt first, and which ones they adopt in sequence later. This helps us and farmer-promoters to best understand appropriate entry points with newly involved communities and farmers. A survey carried out with 72 households in the Bilanga district provides some insights. Most households, when presented with a "basket of possible technologies," seem to select those "foundational" innovations that provide the highest benefit at the lowest cost and make it possible for other innovations to have an impact. For example, farmers need to prevent soil erosion before they can invest in improving its fertility or diversifying crops. In the first year, most households opted for contour rock bunds and organic manure. Agroforestry (FMNR), on the other hand, is more

Table 8.1 Adoption rates of agroecological practices (2014)

Agroecological practice	Number of households adopting	Percentage of households adopting
Rotation*	1,078	52.9
Agroforestry (FMNR)	1,066	52.4
Rock bunds on the contour	816	40.1
Zai planting pits (micro catchments)	406	19.9
Organic manure/compost pits	291	14.3
Improved short cycle seeds	282	13.9
Inter-cropping	121	5.9
Half moon water catchments	42	2.1
Mechanized Zai	5	0.2
Total households n = 2,036 in 15 villages		

*This combines both traditional and new techniques related to crop rotation

Table 8.2 Patterns in the combination of agroecological practices adopted by small scale farmers in bilanga district of Burkina Faso (2014) (72 households surveyed)

Combination of agroecological practices	Number of Households	Percentage of Households
Contour Rock Bunds + Organic Manure	32	82
Contour Rock Bunds + Agroforestry (FMNR) + Organic Manure	23	59
Contour Rock Bunds + Zaï (includes Organic Manure)	20	51
Zaï (includes Organic Manure)	16	41
Contour Rock Bunds + Agroforestry (FMNR) + Zaï (includes Organic Manure)	13	33
Contour Rock Bunds +Short Cycle Seed +Rotation	12	31
Contour Rock Bunds + Agroforestry (FMNR) + Zai (includes Organic Manure) + Short Cycle Seed	11	28
Contour Rock Bunds + Organic Manure + Short Cycle Seed + Rotation	11	28

often applied in the second and third years. The sequencing pattern is also shaped by each household's resources, in particular labor.

In addition to understanding the sequencing of adoption, it is also important to understand how farmers are combining different agroecological practices that they find have synergistic benefits. As presented in Table 8.2, ten different combinations have emerged among the same farm households in Bilanga, but contour rock bunds and organic manure are the most prevalent combination, as the manure is ineffective if washed away by erosion. Moreover, the stone bunds help retain water in the fields so it can penetrate into the soil, providing longer-term moisture for crops. FMNR, once established after two to three years, is a regenerative system that needs to be managed, but generates multiple benefits once established. *Zai* planting pits, which include the use of organic manure, are highly effective but also require relatively significant labor.

Beyond these "foundational practices," other farmers are experimenting with their own innovations and beginning to share these with others. Creating this sort of continuous, farmer-led process of agricultural innovation is one of the end goals of the project. Tani Lankoandé from Sagadou provides a perfect example. She took it upon herself to find new ways to increase agricultural production in the face of climate change. "It started off with a simple observation," she says. She saw that when fallen leaves from nearby trees are transported by rainfall to areas of the field, the soil becomes richer. "These leaves decompose into humus and make the land fertile and arable" she explains. "So, I would collect these dead leaves and put them into small piles throughout my farm, while making sure to add ash. Ash prevents termites from attacking the piles of leaves and the Harmattan winds from blowing the leaves away."

After testing this method out on part of her land and comparing it to another control area, Tani confirmed the usefulness of the practice. ANSD supported Tani in carrying out additional experiments and introduced her to researchers from INERA. She takes pride in the fact that many other farmers in the village have now also adopted the practice. What's best, she says, is that these techniques can be adopted by farmers without the economic resources to invest in new inputs or tools.

A survey of 64 farmer field schools in 2014 revealed that yields of basic crops produced under agroecological conditions increased by 40-300 percent compared to control plots. Based on these successes, farmers want to expand their use of and experimentation with different agroecological techniques. This is important because in the Sahel there is no one single agroecological technique that, by itself, can reverse soil degradation and declining productivity. Transforming the traditional farm into a highly diversified, sustainable and climate-resilient system will require a process of ongoing agroecological innovation in which households progressively learn and adopt a growing range of agroecological practices. Farmers that are convinced of agroecology's efficacy will be motivated to continue with experimentation and implementation, making the *process* of agroecological innovation sustainable into the future.

To this end, a large success of the project has been the creation of agricultural committees in the 60 engaged villages. Forty-seven of these have developed their own action plans for promoting agroecology. To strengthen the capacity of these committees and village leaders in carrying out these plans and managing ongoing processes of agroecological innovation, ANSD facilitated self-assessments with committee members to understand what capacities they believe

---

Kiribamba Pakouma, an example of women's leadership in agroecology

Because of the project, many more women have also become involved in dry season vegetable gardening for both consumption and sale, as well as implementing improved livestock practices. Additionally, they're using simple methods to process produce for storage and sale (such as through solar drying), and have formed savings and credit groups to support these efforts. Kiribamba Pakouma, for instance, took up vegetable gardening after participating in an ANSD training session. She is part of a women's group whose members support each other with savings and collective work on each other's plots. They also exchange agroecological advice. Kiribamba started by investing only 1,000 CFA (US$2) in seeds and inputs, and with the help of resources provided by the women's group, she now provides much of her family's food. In the last season, she donated 20 percent of her surplus to other families, and sold the rest for US$60; enough to reinvest in the farm and pay off some of her children's school fees. "I am proud to contribute financially to my household's expenses," she says.

---

they needed, and how they assess their own strengths and weakness related to those capacities. ANSD also provided practical training for these community leaders on the use of participatory tools for planning and reporting, defining their roles and responsibilities, managing cash boxes, ensuring cooperation with other groups and actors, and coordinating volunteer farmer trainers. Just as with the technical agroecological skills, leaders trained in organizational management skills used the cascading approach to, in turn, train over 800 other community leaders (43 percent were women).

To promote coordination between village committees and other local groups, ANSD convened workshops at the district-level to create coordinated district action plans. It is hoped that these inter-village networks will work in each of the three districts to promote and accelerate the spread of agroecological practices to overcome soil degradation and hunger.

Additionally, key decision makers have become more active in the spread of agroecology. For example: nine local officials from the environment and agriculture ministries co-facilitated training sessions on FMNR, *zai* planting pits, half-moons, and stone contour bunds; three other environment ministry staff co-facilitated information sessions with communities on the laws and regulations regarding the management of trees; the ministry of the environment cooperated with local public radio to broadcast programs to promote FMNR and other agroecological techniques; and local government staff and religious leaders are supporting FMNR and agroecology within their organizations.

Moving forward: Trust the process, build capacity, enable environments

The most critical point to the ANSD approach to scaling agroecology is that it does not involve the transfer of pre-determined packages of technologies. Instead, ANSD works with farmers to identify a "basket" of potential innovations, fosters farmer experimentation and exchange, and enables

Photo 8.4 A woman watering her onions as part of a dry season vegetable gardening project of a women's group in Bassieri.
Photo Credit: Tsuamba Bourgou

each household to apply the combination of agroecology practices that best suits their circumstances. Through this collaboration, ANSD creates an improved *process* for accelerating agroecological innovation and creating positive synergies. This process has the potential to help the local population reverse the vicious cycle of declining soil fertility and food production, and to regenerate farms and improve families' wellbeing on a regional scale.

Batta reflects:

> "This program has shown the importance of focusing not just on technical work, but also on a scaling strategy and advocacy and efforts to create an *enabling environment* for agroecology. In the process, ANSD has seen the importance of selecting pilot villages, of supporting ongoing farmer experimentation, and in achieving early wins in order to create enthusiasm. Working with "foundational" agroecological techniques that can allow for the sequential and combined adoption of other techniques, in order to continuously "deepen" agroecological understanding and practices has been key. So has the geographic spread and expansion outward to new areas. We knew from the outset that it is important to focus on women's capacity building in agriculture, and create diverse alliances. This has proved essential" (Batta, 2015).

As the project moves forward, ANSD continues to work with farmers and community organizations to eventually reach all 125 villages in the area. While we have already witnessed significant tangible change, we estimate that it will take between six and ten more years to truly create widespread, sustainable farming systems at the level of the three districts. Each new wave of agroecological innovations can build on the previous ones, as long as there is strong social organization in place to lead the process. Farmers have identified increased livestock-farming integration and improved biological pest management as next steps for their agroecological innovation.

Farmers like Souobou, profiled in the opening of the chapter, have already made great achievements in their first few years experimenting with and spreading agroecological practices, but they are not content to stop there. Souobou strives not only to increase current productivity, but also to make his farm resilient in the face of climate change and to teach agroecological approaches to others. "In 2013, when rainfall was scarce, many farmers growing maize had very poor harvests," he explains. "I was one of the few farmers with a good maize harvest. I was able to help neighbors and family with food during the lean season. Before adopting the agroecological techniques, I cultivated six hectares. Now I am producing twice as much food, but I only cultivate about four hectares. I have many plans to improve my farm. I am already expanding the stone contour bunds to cover more of my farm. I am also beginning to implement the *zai* technique (micro-water catchments). I will invest in some small tools, like a cart to move stones, and will invest in more livestock for manure composting. These agroecological practices are highly relevant. I am proud that I have learned them, and I am ready to help other relatives and members of my village who are ready to learn."[2]

Notes

1. Adjima Thiombiano. Personal Interview with Amy Montalvo, June 2014.
2. Souobou Tiguidanlam. Interview by Tsuamba Bourgou, June 2014.

BURKINA FASO

BENIN

GHANA

TOGO

CÔTE D'IVOIRE

Bolgatanga

Tamale

Bole

Kumasi

ACCRA

Tarkwa
Cape Coast

GULF OF GUINEA

Regions referenced in chapter

CHAPTER 9
From community to national agroecology movements in Ghana

Bernard Guri and Daniel Banuoko

Summary: *While certain government policies are promoting conventional, monocrop production in the South, farmers in Ghana's Northern Region are building on traditional culture to develop locally situated, agroecological responses to food insecurity and ecological crisis. In only two years, the Center for Indigenous Knowledge and Organizational Development (CIKOD) has created a structure to support these farmers and promote agroecological exchanges. By working with the media, traditional leaders, government agencies, and other associations, they are helping to promote broader knowledge of agroecology (specifically agroforestry) nationally and pushing for more supportive policies and programs.*

Abubakar Sadique Haruna is a farmer in Ghana's Northern Region that was once the country's breadbasket for production of cereals and tubers. Now, the region suffers from growing hunger. Abubakar is also an agro-input dealer. With the help of the Agricultural Development and Value Chain Enhancement Programme (ADVANCE), funded by US-AID (United States Agency for International Development), Haruna hires out his tractor for plowing services to about 400 local farmers; supplies them with improved seeds, agrochemicals and fertilizers; and educates them on practices to increase yields with these inputs. For every acre of land ploughed, these farmers either pay him in cash or in kind with an 84-kilogram bag of maize at the end of the farming season.

"The unfortunate thing," says Haruna, "is that some farmers, after paying for plowing, are not able to afford these agro-chemicals (because of the bad harvests)." The year 2011 was particularly bad, as 200 of his clients had so depleted their assets that they could not even afford to plow their fields. The evaluation officer of the Ministry of Food and Agriculture (MOFA) in Northern Region, Festus Aaron Langkuu, concluded, "Although the government is supporting some farmers with fertilizers, the bottom line is that if there are no rains, these farmers cannot grow their crops, and this will derail (progress towards) the objective of reducing poverty" (Oppong-Ansah, 2012).

Government and development policy: Supporting a "new green revolution," ignoring agroecology

While the government of Ghana has made important progress reducing hunger at a national level, improvements are concentrated in the agriculturally rich areas of the south. Ghana's northern savannah regions, where agriculture is dominated by subsistence smallholder production, shares characteristics with other dryland areas of countries in the Sahel. A 2012 World Food Program report showed that 22.3 percent of the region's population—more than 680,000 people—faced food insecurity, with 140,000 individuals classified as severely food insecure (Hjelm and Dasori, 2012). Maize production in the Northern Region fell over 50 percent in nine years, from 164,200 metric tons in 1991, to 78,800 metric tons in 2000 (Oppong-Ansah, 2012). Chronic malnutrition in the region caused the highest national rate of stunting for children under five, at 33 percent (Ghana Statistical Service, 2015: 155). These children affected by stunting will likely suffer permanent negative effects on their mental and physical capacity for the rest of their lives.

The principal causes for this growing crisis are increasing land pressure and a reduction in the traditional fallowing, leading to collapsing soil fertility (as described in Chapter 6 in a West African context). Climate change is also affecting rainfall patterns. While effective agroecological alternatives exist to address declining soil fertility and increased vulnerability to climate change, the government and many aid agencies invest almost nothing in these. Instead, they typically support external input technology packages and services, such as those distributed by Abubakar Sadique Haruna

The stated national objectives of Ghanaian agricultural policy are to enhance food security through increased productivity, create rural employment, increase agricultural export earnings, and reduce risks in agricultural production and marketing. In practice, the government implements this by prioritizing the promotion of export crops, particularly cocoa, and large-scale commercial farming in the rainfall abundant areas of the south. There is a strong push to "modernize" agriculture through a "new green revolution" approach by delivering technology packages to farmers. These include improved seed varieties, subsidized chemical fertilizers, tractor services, and chemical pesticides and herbicides. In 2012, the Ministry of Food and Agriculture (MOFA) spent a full 46 percent of the country's entire agricultural budget on subsidies for chemical fertilizers, mostly for larger scale farming in the south, which they suggest will generate a higher return on investment.

Ghana has also signed onto the G8 New Alliance for Food Security and Nutrition, agreeing to adopt a legal framework that favors the commercialization of African agriculture and makes land available to foreign investors and multinational companies. In addition, the government and international allies have formulated a Plant Breeders' Bill and Seed Law to privatize the ownership of seeds, commercialize their production, support the introduction of GMO seeds, and erode farmers' rights and traditional roles in improving, saving and reproducing seeds.

Even in the North, which is dominated by peasant farmers, governmental and international development programs apply this same "new green revolution" logic. Most MOFA staff are involved in promoting technology packages to small-scale farmers, and know very little about agroecological farming. "Even when efforts are made to address the growing environmental crisis," says Bern Guri (one of the authors of this chapter), "they are often misdirected, and don't take into account endogenous strategies. The Savannah Accelerated Development Authority's (SADA) program to regreen the north through tree planting is a case in point. Instead of working with communities and promoting native species, the program cleared land and promoted non-native species. The initiative was a failure, with most tree seedlings dying from drought, fires, and animals. These strategies are not working" (Guri, 2015).

Chemical fertilizers can, of course, create short-term production increases. However, as Abubakar's experience reveals, many farmers cannot afford them, so they do not all benefit from government programs promoting them. For those who can, the programs provide a disincentive to transition to more agroecological forms of soil management. Thus, rather than addressing the crisis in the north, Ghana's current policies and programs are instead contributing to increasing farmer debt, vulnerability to climate change, and inequality among many smallholder farmers. In contrast, if they supported farmers' agroecological strategies, the region would see more sustainable increases in productivity and local communities would receive more overall benefits.

Responding to crisis: The Center for Indigenous Knowledge and Organizational Development

Traditional chiefs and authorities play an important role in Ghana's culture and social structure. In June of 2015, Paramount Chief Naa Puowelleh Karbo of Lawra, in Upper West, Ghana, addressed the National Forum on Desertification. The audience included the Ministers of Food and Agriculture; Environment; and Science and Innovation. "Based on what is happening in the communities of Lawra, I urge you to support farmer-managed natural regeneration of trees (FMNR). This form of agroforestry is a crucial strategy to restore soil fertility and food production in our region," he urged.[1]

The Center for Indigenous Knowledge and Organizational Development (CIKOD), founded in 2003, has been working to integrate traditional culture and capacities—such as those that FMNR is based on—into development approaches. CIKOD is a Ghanaian NGO that promotes endogenous (locally generated) development, building upon local assets while integrating appropriate external resources in order to strengthen communities. We believe that this will enable them to improve food production, health, natural resource management, and traditional governance.

Photo 9.1 Women with pruned tree branches from FMNR, to be used for fodder and fuel wood. Eremon, upper west Ghana.
Photo Credit: Daniel Banouko

In 2013, building upon its existing programs in the Lawra and Nandom Districts in the Upper West Region, CIKOD initiated a program with the support of Groundswell International to more systematically address the growing crisis of environmental degradation and malnutrition. Our goals were to strengthen farmer-led innovation and spread agroecological approaches; to foster a broad based movement of small-scale farmers' organizations testing and spreading agroecology within Ghana; and to document and leverage this grounded work through wider campaigns to create more supportive policies locally and nationally. In carrying out this work, we also prioritized the involvement and leadership of women farmers.

Given that the program is only two years old, more time is required to document the adoption of agroecological techniques and measure the impacts on farmers' lives. However, the relatively modest scale of the initial community-level work has been leveraged to build a powerful social movement in Ghana for food sovereignty and farmers rights and against the Plant Breeders' Bill.

Spreading agroecology horizontally

Our community-level work focused on 34 villages, clustered in four traditional clan groupings (Tanchara, Gbengbe, Ko, and Sibr Tang), within Nandom and

Lawra Districts. As with all of its programs, CIKOD prioritized engaging and building trust with traditional authorities, including the chiefs, community leaders, queen-mothers, and the *tingandem* (land priests). These respected local leaders facilitated collaboration with communities and helped coordinate activities to promote agroecology. This helped to accelerate farmer experimentation and learning and the rapid adoption of the improved agroecology practices, particularly FMNR (agroforestry).

"Early on, we showed community leaders a video on the 'Niger story'," reported Daniel Banuoko, "They were shocked by the vision of what could happen in their own area if current deforestation and erosion trends continue. But then they also saw the incredible response of communities in Niger to promote FMNR, restore their soil fertility, and generate many benefits for themselves. After that, we worked with farmers to carry out a participatory assessment of the deforestation and natural resources trends in their communities, and to analyze and compare their practices for managing their farms and trees, and which were beneficial or harmful. They committed to keep what had happened in Niger from happening in Upper West Ghana."

CIKOD and Groundswell International also worked together to organize cross visits for community leaders, traditional authorities, and local government representatives. During these visits, they were able to learn from other rural communities in Ghana and Burkina Faso who were facing similar or more difficult challenges, but were implementing effective agroecological practices. "I was one of the lucky ones selected by CIKOD to go to Bolga (Upper East Region, Ghana) to see FMNR fields in Bongo in 2013," said Lagti Gyellepuo, a farmer from Tanchara who became a leading volunteer tree promoter in his area. "I returned very inspired because I realized that we had a better opportunity than the Bongo people. We had more stumps and shrubs than they have over there. There was therefore no reason for us to fail".[2]

During the study visits, participants identified a set of agroecological practices to test and, if deemed successful, to spread throughout the Upper West Region. These included: farmer-managed natural regeneration of trees (FMNR) to improve soil fertility; soil and water conservation, including the use of soil bunds and tied ridging (which maintains and channels water in a given area by connecting bunds – see Figure 9.1); the use of legume cover crops; intercropping of maize/millet/sorghum; promoting the use of local seeds to diversify production; and strengthening local markets and food systems. "We heard from local people of the benefits of FMNR, which included increased food security and nutrition; more fodder for livestock leading to higher livestock productivity and survival; and women now [having] fuelwood available throughout the year," reported Juliana Toboyee, of the CIKOD staff. "The pride and happiness of the community members was equally evident. What was once a burnt-out, barren hill is now a forested area

Tied Ridging

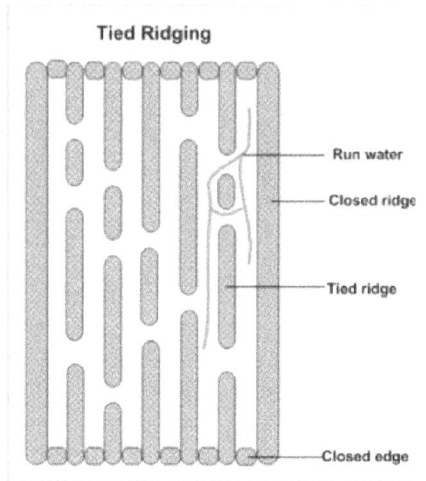

Figure 9.1 Tied ridges

with an abundance of fruits, leaves and tubers, firewood, fodder for livestock, and wildlife" (Toboyee, 2013).

During the learning trips they also identify key factors that had allowed FMNR and other sustainable practices to spread. These included the support of traditional authorities, community ownership of the initiatives and benefits, and community-generated by-laws for managing vegetation and fires.

After a trip, Lagti Gyellepuo explained, "I immediately worked to convince other people in my community and started training them as a part of the CIKOD program. Today my farm is different. Tomorrow my entire community will be different."

CIKOD also worked with communities to identify traditional practices of farming and land management that reflected agroecological principles. They also analyzed how farmers understood the logic of these practices. Based on this understanding, they worked together to improve these practices and to spread them more rapidly to address the growing crisis.

For instance, farmers identified the local term *"tiru guollu"* for their traditional practices for agroforestry and managing the regeneration of trees, and discussed their understanding of what the term meant and what practices it involved. They then discussed and experimented with how they might improve *tiru guollu*, drawing lessons from other FMNR experiences in order to maximize improvements to soil fertility and production of crops, fodder, fruits, nuts, and wood for fuel and building. Key improved practices included greatly increasing the density of trees that could be managed on farms while producing food crops (something they were not used to doing); pruning these trees in ways that supports this co-existence of crops and trees; using the

Photo 9.2 Farmer field visit to a productive agroecological farm with FMNR, upper west, Ghana. *Photo Credit*: Daniel Banouko

organic matter and firewood generated in farming and household practices; and learning the uses and values of different tree species.

Communities recruited farmers as Volunteer Tree Promoters (VTPs), prioritizing women and men who were interested in experimentation, had access to land to do so, and were motivated to lead farmer-to-farmer training. They organized famer-to-farmer exchanges and field days, during which groups of visiting farmers could learn from the experiences of others. This led to the creation of a network of farmer-to-farmer leaders and volunteers promoting agroecology and FMNR across both Lawra and Nandom districts.

We also used a number of other strategies to strengthen endogenous capacity for promoting agroecology. VTPs used exhibitions and shared information on effective principles and practices at the traditional community festivals in Kobine and Kakube. The festivals are widely attended and create an opportunity to quickly reach a great mass of people, including from communities not directly involved in the program. CIKOD also worked closely with the Rural Women Farmers Association of Ghana (RUWFAG), which has over 5,000 members in the Upper West, to enable them to educate and organize other women in their communities. We supported young people to form FMNR clubs in schools, carry out tree planting, and create educational songs and drama sketches to share during the festivals. Our program helped to strengthen traditional governance mechanisms to control bush fires. We at CIKOD also worked with local leaders to help produce local language radio broadcasts to inform community members about agroecological techniques and stimulate discussions through phone-ins. Finally, we used radio and community meetings to promote competitions to give prizes to the best agroecological farmers.

The results of horizontal spread

After the initial two years of the program, the collaboration between CIKOD and rural communities to promote agroecology is already yielding significant results. The VTPs have developed their own songs on *tiru guollu*, which they usually sing while doing group work to prune trees. The most popular song is *"Tikon sage, ti sagkebo,"* which translates to: "We will not allow our land to be degraded. Why should we?" This song has already been integrated into traditional xylophone tunes played at cultural events in all 34 villages. In the second year, farmers organized themselves into four *"Tikon Sage"* ("we won't allow it") groups in the clan areas of Gbengbe, Tanchara,Ko and Sibr Tang. The slogan has become a banner for a growing local movement to promote FMNR in the region, and avoid the degradation that has occurred in other areas.

Additional results of the program include:

- The two Paramount Chiefs plus ten divisional chiefs, who CIKOD educated on the benefits of agroecology and agroforestry, are actively promoting the practices in their traditional clan areas.
- Farmers in two additional villages (beyond the 34 directly involved in the program) have started on their own to adopt and spread key practices. With their help, and other farmers, we plan to expand the program to 60 villages over the next two years.
- Communities have selected 157 Volunteer Tree Promoters, CIKOD has helped to train them, and they are now actively championing FMNR and other agroecological techniques.
- We predict that we will train an additional 785 farmers (five per VTP) by the third year of the program.
- The program has reached over 1,200 households, with over 89 percent of those farmers testing and adopting improved forms of FMNR and other agroecological practices.
- Farmers have measured a 60 percent increased density of trees on their farmland, with 90 percent of the trees regenerated from existing stumps on their land.
- Traditional authorities supported women to gain access to 50 acres of land, with secure tenure for three to five years, in order to transition the land to improved agroecological management and to produce groundnuts in order to improve their livelihoods.
- Farmers groups have established 43 "slash and mulch" demonstration fields in the 34 villages, and four seed multiplication fields.
- Farmer experimentation has led to ongoing, simple innovations to adapt FMNR and agroecology to the local context.
- A contest was carried out at the end of the 2015 to recognize and celebrate the achievements of the best FMNR farmers.

Farmers' perceptions

CIKOD conducted a survey in 2015 to understand farmers' perceptions of the positive aspects, challenges, and recommendations to improve the agroecology and FMNR work. The farmers expressed the following:

- The techniques have improved the traditional system of tree regeneration. It is a sustainable way of farming to improve yields.
- Biomass from leaves obtained from tree pruning improves soil organic matter, fertility, and yields, especially when used to produce compost.
- There is an increase in soil moisture content due to techniques such as tied ridging, which prevents rainwater runoff.
- There is a gradual improvement in the soil structure and soil living organisms as compared to the slash and burn method.
- Fuel wood and "hunger fruits," like berries, are more available on farms for women. Women have additional sources of income.
- Pruned trees have straight stems which can later be used in building.
- Communities are reducing the indiscriminate felling of trees.

$\bigcirc$ **Farmer testimony**

Amata Domo[3]

"I used to farm without pruning. I uprooted trees found on my farm fields and burnt them. However, after adopting and practicing the FMNR technologies of pruning, there are always straight-stemmed trees on my farm that I can use for building purposes. There is always available fuel wood from the cut-off branches. This saves me time and resources going to look for fuelwood to use at home. I even planted groundnuts on land where I used the leaves from pruning as mulch, and did same on [land] without any mulch application as a test. I was overwhelmed at the result, as the one with the mulch turned out better. I therefore recommended agroforestry and FMNR to a friend who intercropped okra on the pruned field. She was really grateful, as she could easily use the straight-stemmed tree for her building as compared to going around looking for straight trees to buy, saving her time and money."

$\bigcirc$ **Farmer testimony**

Kelle Gregory

"Before the program, my farm was infertile. At harvest time I barely had any yields. However, after adopting and practicing the FMNR technologies, I get more and better yields. (Note: The tree count has increased from 80 to about 300 trees on Gregory's 3.5 acres farm.) I combine it with other traditional techniques like tied ridging. In 2013 I only harvested five bags, but in 2014 I harvested seven, even though the rain was bad. My wife does not have to struggle with fuelwood because I reserve the pruned branches for her. This also prevents her from uprooting or felling trees on the farm for fuel. I share this knowledge of FMNR with my family and with three of my friends who are practicing it on their farms. We have to give thanks to our ancestors for guiding us continuously despite economic challenges. It is still best to hope and do what we can to give our children the environment they deserve to live in" (CIKOD, 2015).

Strategies to create an enabling policy environment for agroecology

We have worked through CIKOD to link community-level work (farmer-to-farmer horizontal scaling) with efforts to create a more supportive institutional context and policies to allow for the vertical scaling of agroecology. There is a strong overlap between these strategies.

In the Ghanian context, the first challenge that CIKOD and rural communities faced was to develop and demonstrate a practical, viable agroecological alternative to "agricultural modernization." Once that was initiated in 36 communities in the Upper West Region, CIKOD, farmers groups, and other allies used this experience to influence other actors within the Lawra and Nandom districts, the other 15 districts in the Upper West Region, and in a national context.

District and Upper West region level

In the two districts and the broader Upper West Region, we employed a number of strategies to create a more enabling environment. By **promoting and communicating traditional cultural knowledge** through traditional festivals and local language radio, we educated a broad public. CIKOD also helped organize awareness-raising **District Assembly meetings** in both districts, with farmer champions providing testimonies and showing videos of community practices. This helped to educate elected officials about the efficacy of FMNR and agroecology, in comparison to the government's "new green revolution" and tree planting initiatives.

We facilitated exercises with community members to develop **institutional maps** of all of the main actors and institutions in the Upper West region involved in agriculture, food security, climate change adaptation, and natural resource management. This helped to identify opportunities and obstacles, and in some cases led to alliances and collaboration. For example, we developed an enthusiastic collaboration with the Ghana Fire Service to provide training to community volunteers to prevent bush burning. The chief fire officer for Nandom said that "CIKOD has given us all the necessary support to be able to train fire brigade groups in the three beneficiary communities of Goziir, Ko and Monyupelle. This training on bush fire prevention has created the enabling environment for FMNR and agroecology to thrive in the district. In terms of awareness, the three communities have been convinced that most bush fires are not caused by dwarfs as is traditionally believed" (Yussif and Moshie-Dayan, 2016).

CIKOD was also invited to join a multi-actor platform led by the research initiative CCAFS (Climate Change and Food Security Initiative), which works to help small-scale farmers adapt to climate change. We helped to finance and organize a consultative workshop with multiple agencies, eight traditional chiefs and women leaders, and local government officials. This helped to harmonize and coordinate NGO interventions in Lawra District into a

---

Testimony

Lawra Naa Puowelw Karbo III, Paramount Chief of Lawra Traditional Area[4]

"I was the chairman of the CCFAS-Lawra District Platform, started in 2014. We have organized meetings and programs especially on sensitizing the 30 communities in the district to climate change ... In order for people to understand climate change, they need to accept the idea and see its impact and the difference it is making, to see that climate change is real and not a joke from some meteorologists. We gave importance to the indigenous roles of the communities, and through this they created by-laws to stop burning and protect the bush, and manage the water system. As a former chief planning officer of Ghana, I realize there is a big gap to fill between the people and the government. The effective implementation and enforcement of local laws can only be achieved if the approach is from the bottom to top, not top to bottom. The missing link is the traditional council role in each community. When communities, with the aid of their traditional authorities, make by-laws for themselves, it will be easier for the Assembly to enforce and adopt them. This can then be carried to the national level."

---

consolidated plan for responding collectively to climate change and food security using agroecological and related approaches. CIKOD also used **media coverage,** establishing good rapport with the various newspaper and radio institutions in the two districts, leading to extensive favorable coverage of the agroecology/FMNR initiative.

Finally, CIKOD prioritized the understanding and support of **traditional leaders**. We organized a workshop for both the chiefs and traditional women leaders in Lawra and Nandom districts, and later on the Regional House of Chiefs, to raise awareness on the agroecology/FMNR initiative and foster a critical perspective about existing agricultural policies and practices. This was followed by a workshop for heads of all 17 District Assemblies in the Upper West Region.

National level

In addition to its work in the Upper West Region, CIKOD provided leadership and support to an emerging national coalition advocating for broader food sovereignty and agroecology in Ghana. A galvanizing factor was the "Plant Breeders' Bill," which was being drafted by the parliament and international allies, and pushed forward for approval with limited public understanding, discussion, or democratic debate. As they came to understand the contents of the Plant Breeders' Bill, farmers became concerned about how it would erode their rights to save and reproduce local seed varieties, and what the impacts would be if it allowed and promoted GMOs (genetically modified organisms) in Ghana. A coalition formed, including civil society organizations, farmers' and women's groups, churches, and scientists that pushed for more time for informed public debate, as required under Ghanaian law. CIKOD collaborated with the coalition through strategy development; education and awareness-raising; and media work. "Overall," reported Bern Guri, "we are seeking not just to oppose these negative directions and false solutions offered by the

Photo 9.3 Local seed varieties at a seed fair organized by CIKOD and the rural women farmers association of Ghana.
Photo Credit: Daniel Banuoko

'new green revolution' technologies and GMOs, but to build and strengthen the positive agricultural system that our country requires. We are focused on a 'pro-agroecology' message, and the benefits this will bring for national wellbeing" (Guri, 2015).

The campaign has included a number of key activities to educate the public and encourage democratic debate. For instance, CIKOD collaborated with the *Daily Graphic,* Ghana's major newspaper, to organize a one-day informational workshop for personnel from over 40 media houses in the capital of Accra. This led to widespread coverage on national television, radio, and newspapers.

The Food Sovereignty Ghana coalition also led a public demonstration on January 28, 2014 asking Parliamentarians not to approve the Plant Breeders' Bill (PBB). This helped to further raise awareness and generate petitions from many other civil society and faith-based organizations. A coalition of ActionAid Ghana, the national Peasant Farmers Association of Ghana (PFAG), and CIKOD received a grant from the funding agency STAR Ghana to facilitate and promote civil society and farmer organization democratic input into the review of the PBB. CIKOD organized a series of workshops with PFAG members in both the southern and northern regions of Ghana to provide further information on FMNR and agroecology, strengthening the involvement of the Rural Women Farmers Association of Ghana (RUWFAG) and their "We are the Solution" campaign.

Civil society groups appeal to Parliament

- Not to pass Plant Breeders' Bill

Scan to listen to Mr Opoku

By Severious Kale-Dery, ACCRA

CIVIL society organisations (CSOs) have repeatedly appealed to parliament not to pass the Plant Breeders' Bill because it would amount to surrendering the country's sovereignty.

They said the bill, as it is now, when passed would give seed breeders unlimited powers to the disadvantage of the smallholder farmers.

Right of Breeders

Section 23 of the Plant Breeders' Bill states "A plant breeder's right shall be independent of any measure taken by the Republic to regulate within Ghana — the production, certification and marketing of material of a variety or the exploitation or exportation of the material."

> "The American Academy of Environmental Medicine has urged physicians to advise all patients to avoid genetically modified food."

Speaking at a day's workshop for members of the Peasant Farmers Association of Ghana on the topic, "The Plant Breeders' Bill - What farmers should know", Mr Yaw Opoku of Food Sovereignty Ghana said the bill was a threat to the livelihood of Ghanaians because "the government has no control over the pricing and type of seeds the seed companies will be selling, and whatever the breeders will be doing."

Picture: Mr Bern Guri (hand raised), the Executive Director for Centre of Indigenous Knowledge and Organisational Development, addressing the peasant farmers at a workshop in Accra. Picture: GABRIEL AHIABOR

"This means that the government will not have any control over these who will bring seeds into the country. It is like putting your head, eyes, mouth and cooking power into somebody's kitchen. It means whatever he wants to do to you, because of your stomach, he will do to you. It means you have no control of yourself," Mr Opoku said.

He said it was disgraceful and shameful that "our parliamentarians will forget everything and pass this law which gives total control over seeds in the hands of foreigners. As a country, we are selling our birthright".

Mr Opoku explained to the farmers that since Ghana had joined the world

Trade Organisation (WTO), it was mandatory for her to pass, among others, the Plant Breeders' Bill to ensure that breeders had intellectual property over the seeds or plants they produced.

He said with the bill, a farmer was not allowed to replant, propagate, sell or market without the breeder's consent.

Contributing, the Executive Director of the Centre for Indigenous Knowledge and Organisational Development (CIKOD), Mr Bern Guri, said the Plant Breeders' Bill was at the consideration stage in Parliament, yet there was no opportunity for small-scale farmers to have their voice heard

on the bill.

The workshop was, therefore, an opportunity for the farmers to understand what the bill sought to do, so that they would not be left in the dark.

He said the bill would lead to a take-over of the seed business by corporate seed and chemical companies, a development which would result in the loss of jobs by small scale farmers.

Small-scale farmers

For his part, the Deputy Chief Executive of Food Sovereignty Ghana, Mr Duke Tagoe, said with the bill, small-scale farmers would become labourers to commercial farmers, since they would not be able to compete with commercial farmers.

He said the idea was to edge out small-scale farmers who held the key to the livelihood of the Ghanaians.

Writer's Email: severious.dery@graphic.com.gh

Figure 9.2 Media coverage helped to inform the public on the debate over the plant breeders' bill and the proposed introduction of GMOs into Ghana. *Source*: The Graphic, March 25, 2014

"Previously, even though we were both focused on smallholder agriculture, we did not have a relationship of collaboration with PFAG," said Bern Guri. "We have come to build a vital alliance and collaboration, as we learn from each other and work towards common goals. No one sector or organization can do it alone. We are finding this collaboration with Peasant Association of Ghana, as well as the Rural Women Farmers Association of Ghana, to be vital strategies in creating a broader movement for positive change" (Guri, 2015).

Additionally, CIKOD worked with the highly respected Dr. Kofi Boa to document and disseminate a report on his experimental trials with agroecological approaches and conservation agriculture in southern Ghana, highlighting the effectiveness of these strategies, and helping to promote them to even more people.

Results of efforts to create an enabling context

Though it is a challenging struggle, the community level work supporting farmers to experiment with and spread FMNR and agroecological practices, as well the wider campaign work to educate the broader public and create supportive policies, has had positive results at the local and district level, as well as nationally.

District and Upper West region level

Throughout the Upper West Region, this work has raised awareness, built a growing coalition of allies, and begun to influence and create more supportive design and implementation of programs, plans, and policies. Farmers, women, citizens, and traditional leaders have developed more effective ways to engage. CIKOD was invited by the Environmental Protection Agency (EPA) to share community level strategies and outcomes of the FMNR/agroecology initiative at a National Environmental Day event in Lawra. Obtaining adequate financial resources remains a challenge, but both Lawra and Nandom districts—covering all 34 villages originally involved in the program—are now including FMNR and agroecology in their development plans and budgets. All 34 villages have created village fire brigades and are strengthening traditional community by-laws to manage the burning of bush.

In 2015, the Upper West Regional government invited CIKOD to participate in a regional planning workshop. As a result, FMNR was successfully inserted into the Upper West Region's five-year food security and development plan as a key strategy. This mandate has enabled CIKOD and other allies to promote FMNR within the 17 lower level district assemblies' development plans and budgets. Influencing local development plans and ensuring inclusion in budgets are key steps forward.

After the original successes sharing FMNR and agroecology technologies at the Kobine and Kakube traditional festivals, the two Paramount Chiefs for the districts have requested that these exhibitions be included in the festivals each year. CIKOD has also collaborated closely with RUWFAG to integrate FMNR and agroecology into their work in Lawra and Nandom districts. "It is crucial to work with women," said Bern Guri, "because they have major responsibilities for agriculture, but are also often involved in deforestation through unsustainable harvesting of fuelwood. They need this for household use as well as for income generation, and they generally have lacked viable alternatives. At the same time, women are key to feeding their families. Through the program, RUWFAG now has 133 Voluntary Tree Promoters who are educating other women on FMNR. We are working together to develop alternative income generating strategies, like raising small livestock, vegetable gardening, and strengthening savings and credit groups, so women have alternatives to cutting trees" (Guri, 2015). The program is also working with young people. Over 1,800 school children established the Tanchara Youth Federation, which is educating and promoting FMNR through poetry, drama, and the protection and planting of trees.

National level

At the national level, the campaign strategy sent a strong message to parliament that the public was aware of and concerned about the Plant Breeders' Bill, and the directions it would lead Ghana. This forced the Speaker of Parliament to suspend discussions on the bill for the time being in 2015, and to request the responsible parliamentary committees to engage in further consultation with civil society and the public before reconsidering its passage. Although Ghana was one of the first African countries to sign the African Regional Intellectual Property Organization (ARIPO) protocol to "harmonize" seed laws in Arusha, Tanzania in 2016, as of September 2016 Ghana's parliament had still not carried out the required debate and ratification as a result of strong protests from civil society organizations and peasant farmers. CIKOD, the Peasant Farmers Association of Ghana, ActionAid, and other farmers' groups, women's groups, and civil society organizations have launched a national platform in support of agroecological farming. This will create an ongoing space for dialogue with key national decision makers on how to promote agroecological farming and address concerns on issues such as the Biosafety Act for introduction of GMOs; the Plant Breeders' Bill; and Ghana's involvement in the G8 New Alliance for Food Security and Nutrition.

"We are seeing some important steps forward which we did not initially envision," said Bern Guri. "When we first began dialogue with PFAG, they were advocating for increased government subsidies for chemical fertilizers for peasant farmers. This is the program and strategy they had become accustomed to. Through dialogue and learning together, they are now focusing

Photo 9.4 Protest against the plant breeders' bill.
Photo credit: Daniel Banuoko

on spreading more agroecological approaches to managing soil fertility and are not promoting subsidies of chemical fertilizers. As we have discussed the issue of subsidies for chemical fertilizers at higher national levels, the government has agreed to provide some subsidies to organic compost and fertilizers. In 2016, the national waste disposal company, Zoomlion, began a business with this support to separate organic waste, produce organic fertilizer, and transport it to be used on farmers' fields in the Upper West and elsewhere" (Guri, 2015).

Next steps and lessons for agroecological regeneration

Though this work is quite new, important strides have been made in developing strategies for the spread of agroecological farming in the Upper West, and in contributing to district- and national-level coalitions to more effectively implement current programs and create more supportive policies and alliances for agroecology. The policy environment in Ghana continues to pose major constraints to scaling agroecology, yet this work has demonstrated that it is possible in just a few years to effectively link local (horizontal) and national (vertical) strategies to leverage positive change. CIKOD will continue to work with farmers and women's groups and key allies, such as Groundswell International, in continuing to develop these strategies and expand on this initial progress.

Our experience reveals a number of key lessons and elements of success, which resonate with experiences of promoting agroecology and positive local development around the world. First, it is crucial to **strengthen endogenous development** by valuing and building upon traditional, local culture, knowledge, and leadership. Our program works with farmers to understand their traditional farming methods from their own perspectives. These are often based on agroecological principles. It is important to understand the terms and language farmers have used to describe these methods, and to integrate the traditional understanding and new lessons and practices into cultural expression through songs, theater, local language radio, and traditional festivals. In seeking to improve agroecological farming strategies to face the local crisis and challenges, an important initial step is to **identify effective agroecological practices** appropriate to the context. We have done this by learning from existing innovative farmers engaged in agroecological farming practices, in similar ecosystems and facing similar challenges, through information sharing and learning visits to nearby villages, districts, and even neighboring countries.

A crucial stage after this identification is the **farmer-to-farmer spread** of effective strategies. In doing so, it is important to include traditional leaders and government officials; to train local volunteers; and to focus on women and youth in leadership roles.

Further, we recognize how the *decentralization* of power and decision making to local levels in Ghana has created valuable opportunities for citizen engagement, which means that farmers and women's groups have space for developing viable alternatives and proposals.

Finally, we have learned a number of lessons about building wider alliances and movements. First, it was important to **identify and map actors** who influence agriculture and food systems locally and nationally, and to understand constraints and opportunities for forming partnerships and helping allies to integrate agroecology into their wider mandates and programs. **Developing strong collaborative relationships with media** is important for educating both them and the general public. Also, working to convene decision makers can provide platforms and opportunities for farmers and civil society organizations to inform and share their perspectives.

In working to build wider **social movements** for food sovereignty and agroecology, we are collaborating with broader circle of actors and engaging in the sometimes-difficult work of building wide-reaching alliances. Farmers such as Abubakar Sadique Haruna, the input dealer who introduced this chapter, as well as those from PFAG, RUWFAG, our traditional Chiefs and Women's leaders, government officials, and a wide range of civil society actors, may not each be promoting agroecology yet, but they share common challenges and opportunities for building the Ghana we desire for the future. We will build on the wisdom and tradition of our ancestors, while working to create viable and life-giving alternatives to meet the crises of today, and for our children tomorrow.

Notes

1. As quoted by Daniel Banuoko during the National Desertification Forum for the Northern Region, June 17, 2015.
2. Lagti Gyellpuo. Interview by *CIKOD*, July 11, 2014.
3. Amata Domo. Interview with Daniel Banuoko, June 10, 2014.
4. Lawkra Naa Puowelw Karbo III. Interview with Daniel Banuoku, November 2014.

NORTH SEA

Gronigen

Den Helder

AMSTERDAM

NETHERLANDS

Leiden

Den Haag

GERMANY

Middelburg

BELGIUM

Regions referenced in chapter

CHAPTER 10

Closed-Loop farming and cooperative innovation in Netherlands' Northern Frisian Woodlands

Leonardo van den Berg, Henk Kieft, and Attje Meekma

Summary: *In the context of highly industrialized agricultural systems and centralized environmental management regimes, dairy farmers in the Netherlands have created the space to innovate and develop "closed-loop" agroecological systems through a local cooperative structure. By organizing and building alliances with scientists and others, they have been able to innovate more locally appropriate farming and landscape management practices. They documented and spread the approach, and influenced policy at local, national, and European levels.*

The Northern Frisian Woodlands is a region in the north of the Netherlands that covers 50,000 hectares (about 193 square miles). It has a strong cultural identity and its own language. Since the 1990s, dairy farmers here have challenged the industrial farming model through hedgerow conservation practices, the spreading of healthy manure on soil instead of injecting slurry, and other practices designed to maintain the traditional landscape and biodiversity of the region while developing healthy, viable farming systems.

The farmers have been engaged in a process of developing local, agroecological solutions through partnerships between farmers and with university researchers. The process was sparked by a government regulation on manure injection that the farmers in this region did not believe was suited to their particular context. In the process, they have developed a new cooperative management system for agrarian landscapes, become an example for conservation-agricultural cooperation, and pioneered agroecological innovation. As a result, the rural economy of the region is stronger, product qualities are improved, and there is now more trust and cooperation between farmers and other community residents. In addition, recognizing its unique characteristics, the area has even been declared a "National Landscape" by the government.

Feeding the world

After the Second World War, policy and science became the main drivers in transforming the European countryside into what it is today. Agriculture was pushed into a path of industrialization and scale enlargement. The ultimate

Photo 10.1 Managing dairy cattle in the northern frisian woodlands landscape.
Photo credit: www.duurzamestagehub.nl

goal was to increase production, ostensibly to "feed the world." The means to achieve that goal became monocropping, chemical fertilizers, high yielding varieties and breeds, and imported animal feed. This spurred the growth of animal feed manufacturers, chemical fertilizer producers, and companies developing hybridized or genetically modified seeds and agrotoxins. Their influence grew and intertwined with that of policy and science.

Agricultural production increased and Europe witnessed a sharp decline in the number of mixed farms. As acid rain, excesses of manure, and pollution of ground and surface water plagued many parts of the continent throughout the 1970s and 1980s, it became evident that agricultural development had a cost. The European Union (EU) took action by setting up directives to reduce the emission of ammonia, which was responsible for acid rain, biodiversity reduction, and the leaching of nitrate into ground and surface water (Stuiver, 2008). For example, farmers were no longer allowed to spread manure on the land as they had always done, but now had to inject it into the soil.

Unfortunately, these policy responses addressed symptoms rather than root causes of ecological imbalance, and were far removed from the on-the-ground realities of farming. Compliance with the new rules and regulations required farmers to buy expensive machinery and threatened the future of many farms, including those that were relatively less environmentally destructive. Farmers became tangled in a web of restrictive rules, some of which hampered their own potential to innovate sustainable solutions. For many, these strict environmental regulations, coupled with decreasing food prices and increasing input prices, drove them to abandon farming or migrate to other countries.

The beginning: Territorial cooperatives for farmer-led conservation management

Although nature conservation and agriculture have become their own— apparently contradictory—arenas in Dutch society and policy, farmers in the Northern Frisian Woodlands have always seen them as interdependent. Small-scale dairy farmers' fields in this region, located in the north of the Netherlands, are traditionally surrounded by belts of alder trees and embankments of alder, oak, and bush, which farming families have worked collectively over generations to maintain as part of their farming systems. In the late 1980s, new policies declared these hedgerows as acid-sensitive, and severe limitations were placed on the types of agricultural activities that could be carried out near to them.

While some farmers considered removing hedges before the rules came into force, so as to avoid facing restrictions, many others knew that they could simultaneously preserve the characteristic hedgerows while also sustaining their farm operations *if* they were allowed to do so on their own terms, based on their direct agricultural and environmental knowledge. A group of dairy farmers convinced municipal and provincial authorities to have the local hedgerows exempted from the new regulations. In exchange, they promised to maintain and protect the hedges, ponds, alder rows, and sandy roads in the area.

This gave rise to the first two territorial dairy farmer cooperatives in the Netherlands.[1] Four other organizations were formed soon afterwards, and in 2002, the overarching *Noardlike Fryske Wâlden* (NFW, Northern Frisian Woodlands) cooperative was founded. It currently has a membership of more than 1,000 dairy farmers (almost 80 percent of all the dairy farmers in the area), as well as non-farming community members, and manages about 45,000 hectares of land. When they first started, the territorial cooperatives were a unique organizational structure for coordinated conservation *and* production; no other organization in the country was dealing with the integration of both.

The NFW cooperative was able to align with civil society organiza- tions, especially nature conservation organizations, around a plan for two complementary trajectories. One focused on maintaining and improving the landscape and nature in a way that was compatible with good farming practices, while the other trajectory aimed to develop a strategy for sustainable farming. To overcome the new legislative barriers, the cooperative and its new allies negotiated their ideas with the provincial government and jointly developed a detailed ecological and landscape management plan. They were able to obtain exemptions from several regulatory schemes. The result is that farmers are now managing about 80 percent of the natural and other landscape elements in their area. This includes 1,650 km of alder wooded belts and bank rows, 400 ponds, 6,900 hectares of collectively

Photo 10.2 Managing hedgerows as part of farming systems.
Photo credit: Noardlike Fryske Walden

protected areas to support meadow birds, and about 4,000 hectares for geese (Noardlike Fryske Walden, 2014). The biodiversity has grown richer, and the attractive landscapes are opening up new opportunities for rural tourism and recreation. For instance, the cooperative has worked to restore ancient sandy paths as walking trails and bicycle paths.

By placing increasing focus on the integration of nature, landscape, and agriculture, farmers also found ways to strengthen their farming practices. In the words of one farmer:

> "If you manage the landscape well, biodiversity increases. You get, for instance, more grass species, which positively affects the cows' health. And careful maintenance of the tree belts attracts more birds. They eat the insects that destroy the roots of the clumps of grass. So the more birds there are, the less insecticide you need. Nature and landscape management is thus economically advantageous. That is what I learned in the course of time" (de Rooij, 2010).

Better manure for better soils

Manure management has been at the heart of the struggle between Northern Frisian Woodland farmers and mainstream agricultural/conservation policy. As mentioned above, one of the measures the government took in order to reduce ammonia and the leaching of nitrates into ground and surface water was to require farmers to inject slurry from manure into the soil, instead of spreading it across fields as they had traditionally done. The rationale was that injection would limit run-off and ammonia release into the air, thus protecting broader ecological systems. However, dairy farmers in this region were skeptical; with small fields and high groundwater levels in the spring, their land was not suited to the heavy machinery that is required for slurry injection. The nutrients would also be lost into the groundwater, rather than being absorbed into the soil, thus requiring increasing chemical fertilizers to maintain pastures.

Farmers argued that injecting slurry would kill soil life, and that they had a better idea: produce better quality manure. In 1995, the newly formed NFW Cooperative agreed to undertake an experiment with the government to develop alternative methods for reducing nitrogen leaching. However, national political change in 1998 mandated the experiment be qualified as "scientific research" in order for the region to maintain the exemption allowing it to forego slurry injection. To meet this requirement, the cooperative sought collaboration with alternative-minded researchers from Wageningen University (Verhoeven et al, 2003). This resulted in a nutrient management experiment that included 60 farmers and a small group of scientists of various disciplines.

The experiment with Wageningen University produced an unconventional strategy called *kringlooplandbouw*. "Closed-loop farming," as it is called in English (or closed-cycle farming. See Box 10.2), aims to maximize nutrient cycling on the farm (Stuiver, 2008). The starting point of the research was the goal to improve manure quality. NFW farmers gave their cattle more fibrous feeds, such as grass, and less protein, such as soybean concentrates, than was typical of contemporary, industrial farming. They also mixed microbial additives and straw from their pastures with the manure. This produced more

---

Box 10.1 Innovative approaches to learning

In contrast to the technological fixes and measures developed by agronomists and recommended to farmers, the NFW cooperative took on different forms of learning that give the experience, values, and aspirations of the farmers a central role. New knowledge is gained and disseminated with farmers through a wide range of methods, including nature conservation and landscape management courses, and excursions to other farms in and outside of the region. Methods of learning by doing are often combined with small study groups, in which experiences are exchanged and farmers discuss their successes and failures. Another innovative method is farmer-led scientific research. Farmers raise the questions, the research is carried out on their own farms, and results are discussed between farmers and scientists, as well as within the communities.

Much of what is learnt in these "field laboratories" builds on traditional, and often almost "tacit" knowledge. To farmers, regional characteristics, such as belts and embankments of alder trees, have always been a self-evident part of their farms. Knowledge about local crops and cattle breeds has also been passed down through generations as a base for local agrobio-diversity. The NFW territorial cooperative takes advantage of this wealth of knowledge, revalues it, and also creates a system to spread it further among other farmers.

solid and higher quality manure that improved soil functions. The higher carbon/nitrogen ratio resulted in less nitrogen losses to the environment. Special muck-spreaders were also developed that were suitable for small fields. Although farmers reduced their use of chemical fertilizers, their grass yields began to increase due to improvements in soil biology from healthier manure (Verhoeven et al, 2003).

A 2005 survey of dairy farmers in the NFW showed they used 25 percent less fertilizer than their conventional counterparts (Sonneveld et al, 2009). Other studies suggest that these farmers have a higher economic return, because health expenditures for cattle are lower, fertilizer costs are reduced, and milk cows produce for longer periods of time. Although farmers have to invest more time and labor in closed-loop farming systems than in conventional agriculture, many farmers in the NFW believe it is worth the extra effort as they are compensated with more autonomy and well-being (De Boer et al, 2012).

Today, this approach has spread. With many experts and farmers coming to the Northern Frisian Woodlands to learn, the cooperative has taken up an educational role and regularly organizes guided tours and presentations.

Spread and institutionalization

Closed-loop farming has spread beyond the Northern Frisian Woodlands and is currently practiced by 1,000 of the 18,000 dairy farms in the Netherlands

Photo 10.3 Learning visit with representatives of the ministry of economic Affairs.
Photo credit: Noardlike Fryske Walden

Box 10.2 Closed-loop farming

Today, closed-loop farming (*kringlooplandbouw*) encompasses a whole range of agroecological practices that focus on making the best use of local resources. Whereas conventional agronomy divides the farm into separate entities, closed-loop farming takes a circular, agroecological approach emphasizing the integrated management of different parts of the system: soil quality, feed quality, grassland quality, and animal health (see Table 10.1).

For example, in closed-loop farming, cattle are no longer fed high doses of protein. Instead, they are understood to be grazers and ruminants demanding more fiber and energy—meaning more carbon and less nitrogen in their diet. This diet improves the quality of manure, which in turn improves the soil, leading to improved pasture, better herd health, and higher quality milk and meat. Closed-loop farming also leads to lower emissions and less leaching. It helps close the phosphorous cycle, which is important, given that phosphorous reserves are expected to deplete and the price of phosphorous to become very expensive within 50-70 years. The demand for soy to feed livestock, often associated with deforestation and land grabbing in the Global South, is also reduced. Finally, the agroecological strategy creates more beautiful and biodiverse landscapes. Good manure attracts flies, beetles, and larvae that meadow birds feed on.

Table 10.1 *Kringlooplandbouw* compared to conventional farming practices

Principle	Practices	Results
Feed quality and animal health	Production of own fodder crops, using roughage from natural reserves, reducing digestible crude protein content of feed	Less imports of feed; healthier cows; fewer young cattle are kept as cows live longer; improved milk and meat quality
Soil health	Use of light machinery; less plowing; direct sowing in the sod; feeding the fungi and bacteria in the soil with more carbon and less nitrogen	Less compaction, more organic matter, more soil life; prevent mineralization of organic matter, loss of nitrates, and emission of CO_2
Grassland quality	More permanent grassland; integration of herbs in grassland	Improved animal and soil health
Nutrient use efficiency	More frequent application of smaller amounts; dung is separated from urine in the stables; separate application of the liquid fraction and the solid fraction on the land	Less compaction and better soil structure; lower fertilization levels, lower leaching, reduced ammonia emissions (contains more Organic Matter (C) with slower release of minerals)

(Holster et al, 2014). Its principles have been applied in a range of other large projects in five other provinces.

Scaling, in this case, went beyond the horizontal spread of farming practices. As the closed-loop farming approach grew, it has become recognized and institutionalized in a variety of spheres, thus demonstrating successful vertical scaling as well. The conventional farmers union in the Netherlands now also recognizes, promotes, and defends closed-loop farming. As a result, business and advisory services have designed adapted feed ratios and lower mineral fertilizer doses, and many veterinary doctors now look at the carbon/nitrogen

(C/N) metabolism in the cows' stomachs and recommend higher C/N ratios in feed and fodder to improve their health. Researchers support the pioneering farmers at much larger scales than before, and provinces now recognize that this type of farming supports rather than conflicts with environmental conservation, and are considering supporting its expansion (ibid.).

Closed-loop farming also holds economic promise for farmers, as it is increasingly used in the branding of regional products.[2] Dairy processors are considering paying farmers higher prices for milk produced according to closed-loop farming principles (ibid.).

European subsidies for landscape management

Although the NFW dairy farmers are experiencing positive economic results from closed-loop farming, the cooperative is not yet fully remunerated for their work in cooperative, agroecological landscape management. They do receive compensation from the EU and the provincial government for about half the area under their management, but this hardly pays for the time they must spend on these activities. Most of the European subsidies that have been available for nature conservation are allocated to environmental organizations, keeping with the trend among policy makers and mainstream farmer organizations to ignore or marginalize the idea of farmer-managed landscapes. Recently however, this has begun to change. The NFW cooperative, along with three other cooperatives in the Netherlands, negotiated for better financial support, and in 2015 the new Common Agricultural Policy of the European Union (2014–2020) made provisions for rewarding collectives of farmers for services to society.

Lessons for locally grounded innovation

The expansion of the agroecological strategy of closed-loop farming was not simply a matter of promoting a set of technologies for implementation by farmers. Rather, it evolved over time, as farmers themselves experimented and devised solutions in response to local and national challenges that had grown from the expansion of industrialized agriculture (van der Ploeg, 2008). As farmers initially developed solutions to excess manure and pollution of water sources, they gradually came to better understand the positive interactions and synergies between different elements within a closed-loop farming strategy. It is important to note that the farmers created this space for experimentation themselves, initially in opposition to government policies. They did so by mobilizing other farmers, collectively articulating their problems, envisioning a way forward, rooting farming systems in local ecosystems, creating new organizational structures, and convincing authorities that they could reach policy targets if they were allowed to do so in their own way.

Innovations and solutions were built upon farmers' own knowledge, needs, resources, and aspirations. This ensured that innovations were rooted in the local cultural, economic, and ecological context. Scientists contributed through long-term engagement with this learning process, rather than coming up with their own technological fixes. The process has generated a great many innovations, from management of soil, manure, and hedgerows; to improving livestock fodder; to creating a new cooperative structure for integrated management of the conservation of nature and agriculture; to new policy, market and institutional arrangements. Most importantly, the process involved challenging deeply ingrained ideas of how agriculture should work to "feed the world," which had led to a strong dichotomy between nature and agriculture. Farmers were not able to do this alone, but they organized themselves and built alliances, leading to the articulation of a new collective paradigm of closed-loop farming based on agroecological principles. Alliances forged with scientists and other organizations have been important to strengthen the process and leverage it for wider spread and influencing. By documenting the wider value of these practices for society, farmers and scientists strengthened their argument for wider spread of the innovations. Farmers have also built and maintained working relations with regional, national, and international networks, and with university professors who have advocated their cause at ministerial levels.

Now, other farmers, NGOs, and municipalities from outside the Northern Frisian Woodlands region have become inspired by closed-loop farming and begun experimenting on their own. Other European countries like Denmark have started showing huge interest in learning from this experience. The farmer-led cooperative of the Northern Frisian Woodlands has played an important leadership role in the growth of effective agroecological approaches across Europe.

Notes

1. The Dutch names of these associations are: Vereniging Eastermars Lânsdouwe and the Vereniging Agrarisch Natuur en Landschapsonderhoud Achtkarspelen.
2. This can be seen, for example, in another part of the Netherlands in the marketing strategy of the farmer-cheesemaker association CONO.

CONCLUSION

Supporting a groundswell of agroecological innovation

Steve Brescia

At Groundswell International, we work primarily with partner organizations and marginalized rural communities in the Global South. Some of them are featured in this book. In order to draw lessons from a wider set of experiences, we have also included chapters on the work of organizations and allies in other countries, including the US and the Netherlands, both representing contexts of the Global North. One reason for broadening our lens in this way, and for considering lessons and viable solutions from very different contexts, is the increasingly globalized dynamics and impacts of our agricultural and food systems.

But there are profound differences between realities of the Global South and Global North. In the South, smallholder farming communities are often confronted with life and death challenges related to hunger, access to land and water, climate disasters, migration, and dislocation. They often face weak or undemocratic political systems, lack of protection for basic human rights, and sometimes violence and repression. In general, they have a smaller margin for survival and live more vulnerable lives. Given the often denigrated and fragile ecosystems that characterize smallholder farming communities, and the fact that the majority of poverty and hunger in the world is concentrated in these communities, agroecological farming strategies have proven highly effective and appropriate for improving the lives of peasant farmers in these regions.

In the chapters featured from the Global North, farmers and their allies are also facing real problems in their agricultural and food systems, and are responding creatively with technical and institutional innovations. In particular cases from the US, the Netherlands, and also in some similar situations, they generally do so within a middle-class context in liberal democracies.

Political and economic systems tend to be more developed and functional, rights more widely respected, and farmers have the economic resources and flexibility to organize, mobilize, and pursue alternatives in ways that farmers in the Global South often do not. They are, in relative terms, less vulnerable.[1] These differences have implications for farmers working in each context, as well as for the allies and organizations that seek to support them. Yet across the nine different cases and contexts included here, we can also observe some common principles for scaling agroecology with smallholder farmers. We discuss here some of these principles, drawing on the practical lessons and voices of the people involved. This is followed by an appendix of strategies and methodologies for strengthening and scaling agroecology that emerge from the chapters, and that may be adapted to different contexts by farmers' organizations and support organizations.

Starting points

"The challenges we face are that the rain is insufficient and the soil is declining," said Adjima Thiombiano, as quoted in Chapter 7. "Since the soil fertility has declined, the production has also declined. We don't have as many crops as in the past. There are 11 people in my household. Of course we are worried. If you're responsible for others and you don't have enough to eat, you're very worried."

Smallholder farmers, especially the most economically and politically marginalized, are generally required to first attend to their immediate needs. Those include survival; access to adequate food, income, health care, shelter, and education; and the need to sustain their families, communities, and cultures. This often occurs within a traditional world-view that determines what "good living," or "*buen vivir*," as it is called in some places, means to them (Kerssen, 2015). For instance, Elena Tenelma of Ecuador explains: "In each household in our community, we have the native seeds that we have saved from our ancestors. Taking care of our *Pachamama* [Mother Earth] is the most important thing."[2] Like all people, smallholder farmers will change their practices or strategies when they feel it will benefit them to do so. Policies and incentives can shape their decisions. For agroecology to be deepened, adopted widely by more farmers, and taken to scale, family farmers must believe it provides a better alternative. Disincentives and obstacles must be removed, and enabling factors and incentives increased.

The starting points of family farmers vary by their contexts and the conditions in which they live. Their pathways from these starting points to more productive and beneficial agroecological types of farming are often complicated, and are rarely tidy or linear. Exposure to conventional agricultural inputs (hybrid and GMO seeds, chemical fertilizers, pesticides and herbicides, etc.) and to markets varies for smallholders in different contexts. Peasant and indigenous farmers from the Global South often draw on deep historical wells of knowledge of agroecological farming and natural resource management practices. Yet in many cases, they adopt some combination of agroecological *and* conventional practices. Agrochemical inputs have been promoted to family farmers by agricultural ministries, government subsidies, agribusinesses, NGOs, and philanthropies for decades. Others may be returning to farming and seeking to recapture eroded knowledge and practices, for example having recently acquired land, moved back from urban to rural areas, or started farming in urban and peri-urban settings.

Many households combine different strategies within them: women, for example, may produce agroecologically on one plot over which they have more control and that is dedicated to household consumption, while men use conventional practices on a larger plot for basic grains, perhaps engaging in contract farming with mono-cropping and prescribed external inputs. Families often combine farming and off-farm labor, which may include seasonal employment on industrialized agricultural plantations. In each case, farmers and communities must realistically assess their starting points, challenges, and interests, and develop a process to deepen or transition to agroecology in ways that make sense for them.

Within this complex reality, powerful oases of agroecological farming exist, and some key agroecological practices predominate across populations.[3] The challenge is, how to deepen and spread these agroecological principles and practices to substantially improve wellbeing. Doing so requires bridging work at the grassroots level with wider social movements and policy advocacy. It requires building productive alliances between farmers' organizations and social movements, NGOs, scientists, the government, and local businesses. Model farms and working in isolation will not be enough.

A guiding vision

In addition to a critique of the status quo, we need a positive vision to guide us as we seek to spread agroecological solutions and weave better alternatives for present and future generations. Drawing from the experiences of the women and men featured in this book, we can glimpse some shared key elements of this vision.

Nel, a farmer from the semi-arid region of northeastern Brazil, used techniques that he learned as a migrant working in São Paulo to build a better type of cistern to capture rain water when he moved back to his home community. His innovation was effective, less expensive than typical cisterns, and met a need of local people—so it spread. This eventually contributed to a burgeoning movement to build one million cisterns. It has also fed into a new paradigm of "living with the semi-arid region," emphasizing solutions generated by local people rather than those delivered from above. In Haiti, Jean Luis, a farmer from the North Department, has a vision of restoring mountainous land so that people don't have to migrate to dangerous urban conditions. He is working towards that vision by helping to organize a peasant association that links multiple villages in the task of regenerating social bonds, soil, and farming livelihoods. In Mali, the leaders of the *Barahogon* association have a vision of recovering their traditional roles and knowledge to regenerate trees on their farms and fallows, in order to reverse the growing cycle of defor-estation, desertification, and hunger. Their work is now contributing to a wider movement to re-green the Sahel. In the US, Steve Gliessman and Jim

Cochran, a scientist and a farmer, began working to solve pest and disease problems associated with mono-cultivation of strawberries, and developed a vision to gradually work towards more agroecological farming and a more just food system.

Collectively, the cases in this book emphasize the importance of continuing to innovate and move forward toward that vision of a better future—and to do so in a way that places people, family farmers, communities, and the regeneration of natural resources at the center. The vision is based on the principles of local agency, authentic democracy, and equity.

If we are to achieve such a vision in the future, what might it look like in practice?

First, context will matter. People and communities will create their own versions of "good societies" based on local context and culture. Across different contexts, we can observe some common principles and elements. Farmers must continuously innovate around agroecological principles and practices to develop successful strategies. Local knowledge, innovation, and agency must be fostered, rather than displaced. Technological improvement will matter, but in a way that is people-centered, appropriate, and regenerative. Farmer-to-farmer and community-to-community learning and knowledge sharing networks will play key roles in supporting the spread of these principles and practices. Given the fast pace of disruption created by climate change, finding ways to accelerate farmers' normal pace of innovation and spread of agroecology will be important. Scientists, ministries of agriculture, and NGOs must work collaboratively towards these goals alongside farmers, rather than focusing on mandates to deliver standardized technology packages.

Farmers must be supported to continue to improve, save, and distribute diverse, quality, local seed varieties that are the foundation of food production, biodiversity, and resilience to climate change. Soils, forests, and watersheds must be managed in ways that are sustainable and regenerative—rather than extractive. With appropriate support, family farmers can produce and distribute enough diverse and healthy food so that there is no hunger or malnutrition. In doing so, farmers must also earn enough income to meet their needs and pursue their aspirations. Local economies should be strengthened, so rural communities will be places where people can lead healthy and fulfilling lives, and where young people will want to stay.

Markets clearly matter for smallholder farmers, rural communities, and both rural and urban consumers. They should be strengthened based on sound economic, political, and social principles that emphasize decentralized and well-functioning local markets composed of many farmers, food producers, locally rooted businesses (including farmer enterprises and cooperatives), and local consumers. The current trend of the growing concentration of market power and control in a dwindling number of global agrifood corporations clearly contradicts those principles in unhealthy ways.

For this vision to be realized, governments must be continuously made more democratic, accountable to their citizens, and respectful of human rights. Family farmers must be allowed and encouraged to engage as active citizens. Women must be assured equal rights, opportunities, and access to resources. Societies must ensure that they invest in sufficient public goods in the countryside (health, education, infrastructure, etc.) so that rural areas can flourish and provide the food and the sustainable management of natural resources upon which nations depend. In fulfillment of food sovereignty, nations should have the democratic power to make decisions over how to ensure the production of abundant, healthy, and culturally appropriate food for their citizens.

Is such a vision realistic or a fantasy? If it is not realistic, what's the alternative? While we have a long way to go to reach this vision, the reality is that millions of people are working around the world every day to create it.

Signs of progress

There has been a significant and growing recognition over the last 15 years of the need to transition our dysfunctional agricultural and food systems towards sustainable, productive agroecological systems. This is being expressed to some degree in mainstream institutions and global agreements.

In September 2014, the **Food and Agriculture Organization of the United Nations (FAO)** organized the International Symposium on Agroecology for Food Security and Nutrition. The FAO followed this with three regional meetings in 2015. At the Latin America and Caribbean meeting (Brasilia, Brazil, June 2015), participants agreed on the following recommendations to support the transition from the industrialized food system towards agroecology:

> [For] agroecology to improve household incomes and national economies, it is vital to guarantee territorial rights for small-scale farmers ... Public policies should be promoted to boost agroecology and food sovereignty, in the face of climate change, to be defined, implemented, and monitored with the active participation of social movements and civil society groups, while making the necessary resources available. Participants also called for the necessary institutional conditions to restrict monocultures, the use of chemical pesticides, and land concentration, with the aim of increasing agroecological small-scale production in the region. Other calls were for the fostering of territorial dynamics of social innovation and technology by creating and/or strengthening the interdisciplinary core of agroecology with the capacity to link with the processes of education, research, and learning; also, the official recognition of traditional, ancestral and local knowledge and cultural identity as the basis of agroecology. To achieve this, public research institutes should respect and value traditional knowledge, promoting knowledge dialogues in their research programmes.[4]

In September of 2015, a UN Summit of world leaders adopted 17 **Sustainable Development Goals** (SDGs) to be achieved by 2030. The second Goal is: "End hunger, achieve food security and improved nutrition, and promote sustainable agriculture." One of the eight sub-targets is:

> By 2030 ensure sustainable food production systems and implement resilient agricultural practices that increase productivity and production, that help maintain ecosystems, that strengthen capacity for adaptation to climate change, extreme weather, drought, flooding and other disasters, and that progressively improve land and soil quality (UN, 2014).

In December of 2015, the **Paris Climate Agreement** was ratified, with 195 countries adopting the first ever universal, legally binding climate deal, with the aim of limiting global average temperature increase to below 1.5 degrees Celsius (UN Framework Convention on Climate Change, 2015). Yet in spite of the fact that agriculture claims nearly half of the world's land (Owen, 2005) and accounts for at least one-third of the world's greenhouse gas emissions (Gilbert, 2012), food and agriculture were left out of the main agreement, which primarily focused on energy and transportation. Nevertheless, the plans of most individual countries to achieve the global agreement *do* include commitments on agriculture. In order to reach the agreement's goals of creating low-carbon economies and net zero emissions as soon as possible, as with the energy sector's transition to renewables, it will be necessary to promote a parallel transition from fossil fuel-dependent agriculture and food systems to agroecological farming and sustainable local food systems which are inherently "renewable."

Key strategies and support roles

A common thread throughout the examples in this book is that they were sparked by some crisis that people were experiencing with their current farming systems. These crises were of various types and magnitude. In each context, people worked to create practical solutions to the challenges they faced. This book highlights different strategies and support roles for working with farming communities to develop these solutions and advance and spread the transition to agroecology.

Agroecology is a practice, a science, and a movement, and the work and strategies from all three streams are vital and interdependent. Similarly, the scaling of agroecology happens at three levels: depth; breadth; and verticality. The examples from this book all highlight two positive forces that are essential to agroecology: the innovative capacity of farmers and consumers, and the regenerative power of our ecosystems. A basic lesson is that effective strategies should work to strengthen and build upon these positive forces, rather than displacing or undermining them. Another is the importance of different actors and organizations working collaboratively and synergistically.

One way to map and understand these inter-weaving roles, and identify where gaps, complementarities, and opportunities may exist, is through a simple matrix that takes into account these categories. In Figure C.1, we highlight a small number of examples from this book of different strategies within each category.

Most initiatives and programs will address some aspects better than others, as is the case with the examples from this book. Such a mapping exercise can be applied to identify gaps, opportunities for collaboration, and synergies at territorial, national, or regional levels.

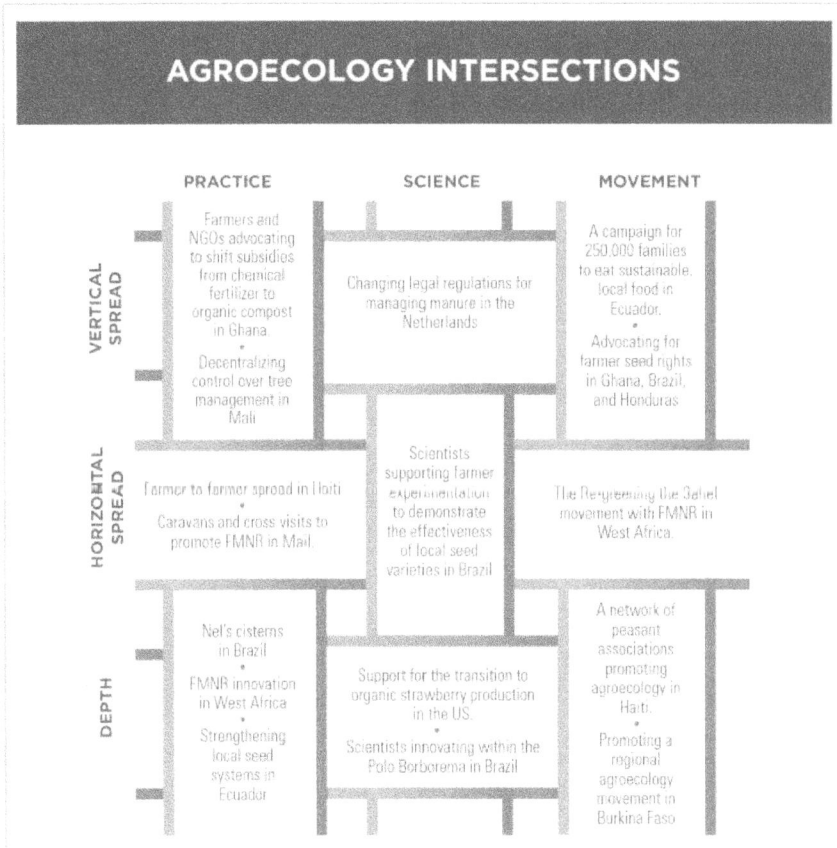

AGROECOLOGY INTERSECTIONS

	PRACTICE	SCIENCE	MOVEMENT
VERTICAL SPREAD	Farmers and NGOs advocating to shift subsidies from chemical fertilizer to organic compost in Ghana. • Decentralizing control over tree management in Mali.	Changing legal regulations for managing manure in the Netherlands.	A campaign for 250,000 families to eat sustainable, local food in Ecuador. • Advocating for farmer seed rights in Ghana, Brazil, and Honduras.
HORIZONTAL SPREAD	Farmer to farmer spread in Haiti. • Caravans and cross visits to promote FMNR in Mali.	Scientists supporting farmer experimentation to demonstrate the effectiveness of local seed varieties in Brazil.	The Regreening the Sahel movement with FMNR in West Africa.
DEPTH	Nel's cisterns in Brazil. • FMNR innovation in West Africa. • Strengthening local seed systems in Ecuador.	Support for the transition to organic strawberry production in the US. • Scientists innovating within the Polo Borborema in Brazil.	A network of peasant associations promoting agroecology in Haiti. • Promoting a regional agroecology movement in Burkina Faso.

Figure C.1 Agroecology intersections.

Weaving the new paradigm

We must continue to weave these strategies and create more agroecological and people-centered farming and food systems from the ground up. The process starts with the actions and innovation of farmers like those featured in this book. It spreads through farmer-to-farmer and farmer-to-consumer movements. Appropriate and enabling policies are crucial for agroecology to thrive. Yet as described by agroecology leaders like Pacho Gangotena of Ecuador (see Chapter 4), even well-intentioned governments can not simply build agroecology from the top-down; "social change in agriculture ... will come from the millions of small farming families that are beginning to transform the entire productive spectrum."

Much like the tenacious tree roots of West Africa's cleared Sahelian landscapes that survive beneath the surface and are now beginning to regrow and heal the land, agroecology has deep historical roots and nourishing stores of wisdom. We still have a long way to go in making this more hopeful vision of the future a reality. The creative power of family farmers, innovating with nature, is a potent and beneficial force in ensuring that we get there. We choose to support that journey.

Notes

1. This is not to ignore the reality of farmers in the Global North who experience high levels of vulnerability and marginalization due to economic marginalization and racial discrimination. Groundswell is in the process of developing strategies on the ground to work in these contexts in the US, yet these are not yet sufficiently developed to include as cases in this book. For a good representation of the agroecological experiences of farmers of color in the United States, see: Bowen, Natasha. *The Color of Food: Stories of Race, Resilience, and Farming.* Gabriola Island, British Colombia, Canada: The New Society Publishers, 2015; and Holt-Giménez, Eric and Yi Wang. "Reform or Transformation? The Pivotal Role of Food Justice in the U.S. Food Movement." Race/Ethnicity: Multidiscliplinary Global Contexts, 5(2011):83-102
2. Elena Tenelema. Interview with EkoRural, 2012.
3. For example, saving and improving local seed varieties, or intercropping.
4. As described by TWN. 2014. "FAO Regional Meetings on Agroecology Call for Policy Change to Support Transition." http://www.twn.my/title2/susagri/ 2016/sa507.htm.

Some strategies and methodologies for strengthening and scaling agroecology

This appendix summarizes some strategies and methodologies, drawn from the cases in this book, that can be adapted, used, and improved to deepen and spread agroecology. It is hoped that these may contribute to the work of farmers' organizations, social movements, NGOs, agricultural ministries, and international development and funding agencies interested in these goals.

Depth

How can we support farmers in transitioning from using one or a few agroecological techniques to create more fully developed, agroecological farming systems?

1. **Farmer-led Experimentation:**[1] It is essential to support processes of farmer experimentation in order to develop appropriate, context-specific agroecological techniques and strategies. NGOs, scientists, and government agencies can productively support this through a "dialogue of wisdoms" between traditional knowledge, people's science, and formal science. Some effective strategies:

   - Identify key constraints and limiting factors
   - Support on-farm experimentation on a small part of wider farm plots that does not put households at risk
   - Limit the techniques being tested initially to one or a few complementary strategies, so that factors of success can be identified
   - Allow farmers to use simple methods and tools to assess and compare the results of agroecological innovations with existing strategies
   - Generate rapid, recognizable results that create valuable benefits for households, and enthusiasm and motivation among farmers. The ability to generate success and the motivation of local farmers are essential steps.

2. **Discovery-based learning in groups:** Experimentation is best done as part of group processes, which is more effective for cocreating and sharing knowledge. Building on traditional knowledge, farmers' understanding of agroecological principles can be deepened.

3. **Replacing conventional with agroecological alternatives:** The starting point is farmers' existing practices. Some traditional

smallholder farmer practices are agroecological, while others are not. Common examples of non-sustainable practices in some contexts are slash and burn to clear land for planting, lack of soil and water conservation, or allowing free grazing of livestock that makes it difficult for farmers to extend the season for agroecological production. As alternatives, farmers may integrate crop residues into soils, use cover crops or green manures, use live or physical contour barriers, and produce fodder and pen livestock to increase production and better use manure in compost.

4. **Foundational technologies:** Test and spread foundational technologies that address widespread constraints faced by many farmers. For example, improving soil fertility through conservation barriers and green manures/cover crops, integrating trees into farming systems, or water harvesting. If successful, these technologies may enable the testing and adoption of other practices, such as increased diversification of farming systems.

5. **Ongoing innovation to deepen the transition to diversified agroecological farming systems:** Adopting just one or two techniques is generally not sufficient to ensure regenerative and resilient farming systems. Strategies can be developed to enable small scale farmers in a given agroecological context to progressively make a transition to more agroecological farming systems through an appropriate sequencing and combination of techniques that address soil management, seeds management, water management, biodiversity management, livestock management, post-harvest storage, access to markets, etc. Farmers will make their own decisions, based on their own assessments of their context perceptions of costs and benefits, as to what practices to adopt and when (Uphoff, 2002).

6. **Leadership and capacity development for women and youth, in addition to men:** Supporting innovation and local organizational development to promote agroecology creates opportunities for learning and the development of practical and leadership skills, in particular for those who are traditionally excluded from these opportunities: peasant farmers, indigenous people, women, and youth. Yet it is crucial to develop explicit strategies to include and empower women in agroecological development. This is true due to a number of factors: women bear important responsibilities for agricultural production; they are also generally responsible for feeding and maintaining their families; women often remain connected to their land, families, and communities when men migrate seasonally or otherwise; and in many contexts and cultures women are deprived of power, decision-making ability, and opportunities. Similarly, explicit strategies should be developed for youth, allowing them to create viable futures in rural communities.

7. **Complementary activities can enable success:** Successful deepening, adoption, and spread of agroecology often require complementary activities that work in synergy with the agronomic approaches.

These can include, for example: savings and credit groups to provide resources for and lower the costs of key investments; community seed banks or tool banks to ensure wide access to these resources; water harvesting; grain reserves to allow groups of farmers to reduce dependence on middle men and store grain locally for consumption or sale when prices rise; and community health activities that prevent diseases.

8. **Local organizational capacity:** The above activities generally require local organizational capacity of community-based organizations and farmers' and women's groups. They often need to strengthen their capacity to coordinate farmer-learning process, and to mobilize and manage local resources. In addition to technical agroecological skills, it is important to strengthen organizational capacities.

Horizontal spread

How can we support the spread of agroecological principles and practices to many more farmers and communities?

1. **Farmer-to-farmer spread:** Farmers who have successfully developed their skills for agroecological innovations are the best teachers of other farmers, because they can share their knowledge in their own languages and within their own cultural and ecological contexts. They use the example of their own farms, as well as their knowledge and communications abilities, to teach others. It is difficult for a farmer who is not successfully practicing agroecology on her or his own farm to convince others to do so.

2. **Learning on farms and in communities:** Various strategies exist for farmer-to-farmer learning, such as:

   • Peasant organizations with agroecology promoters. Farmer promoters may be volunteers, or can be compensated by a farmers' organization, or by other farmers, either in kind, through shared labor, or with money.

   • A successful farmer inviting a group of farmers to visit her or his plot; and then providing follow up and trouble-shooting support to those farmers.

   • Farmer field schools, through which a group of farmers systematically engage in action and learning to address challenges, through a process of regular meetings, experimentation, and analysis.

   • Learning cross visits or organized field days, where farmers from different communities gather in another community to visit successful experiences. A participatory dialogue is facilitated for them to exchange ideas, learn lessons, and identify practices they want to test on their own farms and communities.

   • Participatory assessments or evaluations, where multiple stakeholders (representatives of different communities, local agriculture ministry

or government officials, scientists, etc.) jointly analyze an agroecological experience.
- Organizational structures such as inter-village associations, women's groups, village agricultural committees, or savings and credit groups, which can function as spaces for continuous learning.

3. **Working with broader social movements:** Systematic collaboration with existing farmers' and women's associations organized across wider populations can accelerate the learning and spread of agroecological practices.
4. **Geographic and territorial strategies:** In seeking to spread agroecological innovation within a territory that has common agricultural, ecological, and cultural characteristics, it can be valuable to create an intentional strategy for geographic spread. This may include identifying dispersed, "lead" villages that are strategically positioned to then reach and spread effective strategies to a wider grouping of villages around them. Similarly, highly motivated, innovative farmers may be identified within those lead villages as initial experimenters and innovators who can then share with others. This kind of "cascading strategy" can be used by farmers' organizations and support organizations to foster the spread of agroecology in a cost-effective and rapid manner.
5. **Communications:** Alternative means of communications such as local language radio, popular video, community theater, "best farmer" contests, traditional fairs, or community seed fairs can help to spread information and provide motivation.
6. **Strengthening organizational capacity of farmers' organizations and networks:** Strong organizational capacity of farmers' organizations is essential for leading processes for ongoing experimentation, innovation, and spread of agroecological practices. NGOs can be supportive, while avoiding creating dependencies, by negotiating partnership strategies with farmers' organizations to accompany and strengthen their capacities on specific management, methodological, or technical issues, at the level of villages, inter-village associations, or wider networks. Capacity self-assessment tools can help identify areas for support and collaboration.
7. **Critical mass:** If a critical mass of 35-40 percent of farmers in a community can be encouraged through formal, community-led processes to test and adopt agroecological principles and practices, which are seen by farmers as beneficial, a self-spreading multiplier effect often then occurs, through which the majority of interested households are reached.
8. **Enabling the longer-term transition:** A transition period of 1–3 years, and an investment of time and labor, may be required for farmers to see the growing and sustained benefits of agroecological farming. The learning processes and costs associated with this transition can be supported through savings and credit groups to provide access to

credit at low interest rates; traditional shared-work groups (e.g., *kombit* in Haiti, *minga* in Ecuador, etc.) to support labor intensive work, such as constructing contour barriers; community-managed seed banks and tool banks, managed with revolving loan strategies, to reduce the costs and increase the access to these assets; local grain reserves to improve incomes for farmers that are normally captured by intermediaries; and household or community water cisterns or wells. Such strategies can strengthen social capital and allow farmers to mobilize and manage local resources.

Vertical spread

How can we support the creation of an enabling context for agroecology at the level of policies, institutions, and markets?

1. **Building alliances, linking horizontal and vertical spread strategies:** Linking strong, evidence-based, agroecological processes at the farmer-to-farmer and community level, to wider regional or national farmers' organizations, women's organizations or food sovereignty networks advocating for policy change, is mutually reinforcing. Frequently, technical community-level work is not adequately linked to advocacy for enabling policies, or advocacy campaigns are not adequately grounded in the experiences of smallholder farmers. The two are not as effective in isolation.
2. **Documentation:** Documenting the evidence of the effectiveness of agroecological strategies, in comparison with conventional technology packages for example, is an important tool for wider influencing.
3. **Creating enabling policies at the community, territorial, regional, national, and international levels:** It is often easier for community and local peasant associations to first advocate for changes in policies and regulations at their local and territorial levels. This can include using laws that decentralize decision making and budgeting, and negotiating for farmer decision-making and control over the management of land and natural resources. Constructive collaboration and relationships with local government or agricultural ministry officials can be developed. Successful local models can then be documented and leveraged for wider scaling within countries and across borders.
4. **Bringing policy makers and opinion leaders to the field:** Organize "caravans" or multi-stakeholder field visits by policy makers, media and opinion leaders to visit well-developed examples of agroecological farming that demonstrate the potential for impact at scale. Ideally this would include not just technical agricultural issues, but changes in local regulations or government programs that have helped to enable success, and that could be applied elsewhere.

5. **Policies are necessary, but are not enough:** Even when strong enabling policies are created, such as support for food sovereignty and agroecology in Ecuador's constitution, this is not sufficient to ensure the spread of agroecological farming. Proponents of industrialized agriculture will continue to promote their interests, and will often have greater access to lawmakers than do farmers. Even assuming a highly supportive policy environment, agroecology cannot simply be mandated from above, but depends on the continuous agency, innovation, and practices of farmers to develop and spread it in their own context.

6. **Strengthen local markets:** Stronger links to local markets and short value chains can help incentivize agroecological production. This may be supported through government policies that guarantee a market for smallholder farmer agroecological production, such as local purchases for school feeding programs; through farmers developing contracts with local businesses, such as hotels; through participatory certification and labeling processes that identify agroecological products and inform consumers; or through alternative market arrangements, like *canastas comunitarias* (community supported agriculture) in Ecuador, that directly link consumers and family farmers. Creative communications campaigns can build awareness and support among consumers, encouraging them to invest their food budgets in healthy, local food and the rural communities that produce it, rather than in imported, foreign foods that are marketed to consumers as superior, but often are nutritionally inferior.

7. **Reform agricultural universities and agricultural extension programs:** Agronomists and extension agents are often the main technicians that interact with farmers on behalf of governmental and non-governmental agencies, promoting chosen technology packages and farming strategies. Few are trained in agroecological science and principles, or the practices of supporting farmer-led experimentation and farmer-to-farmer spread of innovations. University programs and extension systems should be reformed to create the next generation of practitioners to spread agroecological processes and alternatives.

8. **True cost accounting of food and farming:** Develop policies and pricing that make transparent and reflect the true costs of conventional vs. agroecological farming to societies.

Note

1. For more, see: Bunch, Roland. 1985. *Two Ears of Corn*. Oklahoma City: World Neighbors.

APPENDIX 2
Literature on agroecology

The following is a partial list of the growing body of literature on agroecology published in the English language. Many additional reports, books, and scientific articles have been produced by scientists and practitioners from countries around the world, including those profiled in this table and included in this book.

Useful links (2023)

- Agroecology Fund
- Agroecology Now!
- Alliance for Food Sovereignty in Africa (AFSA)
- Andhra Pradesh Community-Managed Natural Farming (APCNF)
- Biovision
- Coventry University Center for Agroecology, Water and Resilience
- Cultivate!
- ETC Group
- Food and Agriculture Organization (FAO), Agroecology Knowledge Hub
- Global Alliance on the Future of Food
- GRAIN
- Groundswell International
- IPES Food (International Panel of Experts on Sustainable Food Systems)
- La Via Campesina
- McKnight Foundation, Global Collaboration for Resilient Food Systems
- Millennium Institute
- Movimiento Agroecológico de América Latina y el Caribe (MAELA)
- Oakland Institute
- The Transformative Partnership Platform on Agroecology
- University of Vermont Institute for Agroecology

Books

Title	Organization/ Publisher	Author	Date of Publication
Agroecology: The Ecology of Sustainable Food Systems	CRC Press	Gliessman, Stephen R.	2006
Agroecology: The Science of Sustainable Agriculture, 2nd edn.	Westview Press	Altieri, Miguel A.	1995
Two Ears of Corn: A Guide to People-Centered Agricultural Improvement, 3rd edn.	World Neighbors	Bunch, Roland	1995

Reports and articles

Title	Organization/ Publisher	Author	Date of Publication	Link (for e-book)
Agroecology: The Bold Future for Africa	AFSA & TOAM	AFSA	2015	http://afsafrica.org/agroecology-the-bold-future-for-africa/
The Future of Food: Seeds of Resilience, A Compendium of Perspectives on Agricultural Biodiversity from Around the World	Global Alliance for the Future of Food	Frison, Emile et al.	2016	http://futureoffood.org/wp-content/uploads/2016/09/Future_of_Food_Seeds_of_Resilience_Report.pdf
From Uniformity to Diversity: A paradigm shift from industrial agriculture to diversified agroecological systems	International Panel of Experts on Sustainable Food Systems (IPES)	Frison, Emile et al.	2016	http://www.ipes-food.org/images/Reports/UniformityToDiversity_FullReport.pdf
Building, Defending and Strengthening Agroecology: A Global Struggle for Food Sovereignty	Centre for Agroecology, Water and Resilience, ILEIA and Coventry University	Anderson, Colin et al.	2015	http://www.agroecologynow.com/wp-content/uploads/2015/05/Farming-MattersAgroecology-EN.pdf
Agroecology: Putting Food Sovereignty into Action	Why Hunger	Why Hunger	2015	http://www.whyhunger.org/uploads/fileAssets/6ca854_4622aa.pdf
From Vulnerability to Resilience: Agroecology for Sustainable Dryland Management	Planet@Risk	Van Walsum, Edith et al.	2014	https://planet-risk.org/index.php/pr/article/view/46/154
Scaling-Up Agroecological Approaches: What, Why and How	Oxfam-Solidarity Belgium	Parmentier, Stéphane	2014	http://www.fao.org/fileadmin/templates/agphome/scpi/Agroecology/Agroecology_Scaling-up_agroecology_what_why_and_how_-OxfamSol-FINAL.pdf
Family Farmers: Feeding the world, caring for the earth	Food & Agriculture Organization of the United Nations (FAO)	FAO	2014	http://www.fao.org/docrep/019/mj760e/mj760e.pdf

(conntinued)

Title	Organization/Publisher	Author	Date of Publication	Link (for e-book)
The Transnational Institute at Voedsel Anders	Voedsel Anders/Food Otherwise Network	Sandwell, Katie et al.	2014	http://groundswell.wpengine.netdna-cdn.com/wp-content/uploads/va_report_final.pdf
Agroecology: What it is and what it has to offer	IIED	Silici, Laura	2014	http://pubs.iied.org/14629IIED/
Final report: The transformative potential of the right to food	United Nations General Assembly	De Schutter, Olivier	2014	http://www.srfood.org/images/stories/pdf/officialreports/20140310_finalreport_en.pdf
Confronting Crisis: Transforming lives through improved resilience	Concern Worldwide	Concern Worldwide	2013	https://doj19z5hov92o.cloudfront.net/sites/default/files/media/resource/confronting_crisis_resilience_report.pdf
Smallholders, food security and the environment	International Fund for Agricultural Development (IFAD) & United Nations Environmental Programme	International Fund for Agricultural Development (IFAD)	2013	http://groundswell.wpengine.netdna-cdn.com/wp-content/uploads/smallholders_report-1.pdf
The Law of the Seed	Navdanya International	Shiva, Vandana et al.	2013	http://www.navdanya.org/attachments/lawofseed.pdf
Trade and Environment Review 2013: Wake up before it is too late: Make agriculture truly sustainable now for food security in a changing climate	United Nations Conference on Trade and Development (UNCTAD)	Hoffman, Ulrich, et. al.	2013	http://unctad.org/en/publicationslibrary/ditcted2012d3_en.pdf
Agricultural Transition: A different logic	The More and Better Network	Hilmi, Angela	2012	http://www.utviklingsfondet.no/files/uf/documents/Rapporter/Agricultural_Transition_en.pdf

(continued)

Title	Organization/ Publisher	Author	Date of Publication	Link (for e-book)
Nourishing the World Sustainably: Scaling Up Agroecology	Ecumenical Advocacy Alliance	Prove, Peter and Sara Speicher, Editors	2012	http://groundswell.wpengine.netdna-cdn.com/wp-content/uploads/Nourishing-the-World-Sustainably_ScalingUpAgroecology_WEB_-copy.pdf
Seed Freedom: A Global Citizens' Report	Navdanya	Shiva, Vandana et al.	2012	http://www.navdanya.org/attachments/Seed%20Freedom_Revised_8-10-2012.pdf
Ending the Everyday Emergency: Resilience and children in the Sahel	Save the Children, World Vision, and members of the Sahel Working Group	Gubbels, Peter	2012	http://www.wvi.org/agriculture-and-food-security/publication/ending-every-day-emergency
Escaping the Hunger Cycle: Pathways to Resilience in the Sahel	Sahel Working Group	Gubbels, Peter	2011	http://reliefweb.int/sites/reliefweb.int/files/resources/Pathways-to-Resilience-in-the-Sahel.pdf
Smallholder Solutions to Hunger, Poverty and Climate Change	Food First and ActionAid International	Shattuck, Annie and Eric Holt-Giménez	2011	https://foodfirst.org/publication/smallholder-solutions-to-hunger-poverty-and-climate-change/
Agriculture: Investing in Natural Capital	United Nations Environment Programme	Herren, Hans R.	2011	http://web.unep.org/greeneconomy/sites/unep.org.greeneconomy/files/field/image/2.0_agriculture.pdf
Sustainable Intensification of African agriculture	International Journal of Agricultural Sustainability	Pretty, Jules et al.	2011	http://www.tandfonline.com/doi/abs/10.3763/ijas.2010.0583
Report submitted by the Special Rapporteur on the right to food	United Nations General Assembly	De Schutter, Olivier	2010	http://www2.ohchr.org/english/issues/food/docs/A-HRC-16-49.pdf

(continued)

Title	Organization/ Publisher	Author	Date of Publication	Link (for e-book)
Synthesis Report: Agriculture at a Crossroads	International Assessment of Agricultural Knowledge, Science and Technology for Development (IAASTD)	McIntyre, Beverly D. et al.	2009	http://www.unep.org/dewa/agassessment/ reports/IAASTD/EN/Agriculture%20 at%20a%20Crossroads_Synthesis%20 Report%20 (English).pdf
Nyéléni Declaration on Food Sovereignty	Vía Campesina	Vía Campesina	2007	https://viacampesina.org/en/index. php/main-issues-mainmenu-27/ food-sovereignty-and-trade-mainmenu- 38/262-declaration-of-nyi
Agroecological Approaches to Agricultural Development, Background paper for the World Development report 2008	World Bank	Pretty, Jules	2006	https://openknowledge.worldbank. org/bitstream/handle/10986/9044/ WDR2008_0031.pdf;sequence=1
Soil Recuperation In Central America: Sustaining Innovation After Intervention	International Institute for Environment and Development (IIED)	Bunch, Roland and Gabinò López	1995	http://pubs.iied.org/pdfs/6069IIED.pdf

References

Introduction

Altieri, Miguel (1995). *Agroecology: The Science of Sustainable Agriculture.* Boulder CO: Westview Press.

De Schutter, Olivier (2010). "Report Submitted by the Special Rapporteur on the Right to Food." United Nations, December.

Dobbs, Richard, Corinne Sawers, Fraser Thompson, James Manyika, Jonathan Woetzel, Peter Child, Sorcha McKenna, and Angela Spatharou (2014). "How the world could better fight obesity." McKinsey Global Institute, November.

ETC Group (2013). "Twenty things we don't know we don't know about World Food Security." September. Accessed November 7, 2016. http://www.etcgroup.org/sites/www.etcgroup.org/files/Food%20Poster_Design-Sept042013.pdf.

Food and Agriculture Organization of the United Nations (FAO) (2009). "The state of food insecurity in the world." Rome: FAO.

FAO (2014). "Family Farmers: Feeding the world, caring for the earth." 2014. Accessed November 7, 2016. http://www.fao.org/docrep/019/mj760e/mj760e. pdf.

FAO, IFAD, and WFP (2015). "The State of Food Insecurity in the World 2015, Meeting the 2015 international hunger targets: taking stock of uneven progress." Rome: FAO.

Gilbert, Natasha (2012). "One-third of our greenhouse gas emissions come from agriculture." *Nature*, October 31.

Hickel, Jason (2016). "The True Extent of Global Poverty and Hunger: questioning the good news narrative of the Millenium Development Goals." *Third World Quarterly*, 37: 749-767.

IAASTD (n.d.). "Towards Multifunctional Agriculture for Social, Environmental and Economic Sustainability." Accessed November 7, 2016. http://www.unep.org/dewa/agassessment/docs/10505_Multi.pdf

KPMG International (2012). *"Expect the Unexpected: Building business value in a changing world."*

OCHA on Behalf of Regional Humanitarian Partners (2014). "2015 Humanitarian Needs Overview-Sahel Region." December. Accessed November 7, 2016. http://reliefweb.int/report/mali/2015-humanitarian-needs-overview-sahel-region.

Vía Campesina (2007). "Declaration of the Forum for Food Sovereignty." Nyeleni, Mali, February. Accessed November 7, 2016. https://nyeleni.org/spip.php?article290.

WHO (2016). "Obesity and Overweight." Fact Sheet, June. Accessed November 7. http://www.who.int/mediacentre/factsheets/fs311/en/.

Chapter 1

Galindo, W. (ed.) (2013). *Vozes da Convivência com o Semiárido*. Recife, Centro Sabiá; Conti, L.I..

IBGE (2010). Censo Demográfico Brasileiro. Brasília.

Petersen, P.; J.C. Rocha (2003). "Manejo ecológico de recursos hídricos en el semiárido brasileño; lecciones del agreste paraibano". *Leisa: Revista de Agroecologia*. Vol. 19:2.

Petersen, P.; L.M. Silveira; P. Almeida (2002). "Ecossistemas naturais e agroecossistemas tradicionais no agreste da Paraíba: uma analogia socialmente construída e uma oportunidade para a conversão agroecológica". In: Silveira, L.M.; P. Petersen; E. Sabourin, *Agricultura familiar e agroecologia no Semiárido Brasileiro; avanços a partir do agreste da Paraíba*. Rio de Janeiro, AS-PTA. p. 13–122.

Petersen, P.; L. Silveira; E. Dias; A. Santos; F. Curado (2013). Sementes ou grãos; lutas para desconstrução de uma falsa dicotomia. *Agriculturas*. Rio de Janeiro: AS-PTA. v.10:1.

Pontel, E. (2013). "Transição paradigmática na con- vivência com o semiárido". In: Conti, L.I.; E.O. Schoroeder, *Convivência com o semiárido brasileiro; autonomia e protagonismo social*. Brasília, Ed. IABS. p. 21-30.

Sabourin, E. (2002). Manejo da inovação na agricultura familiar do Agreste da Paraíba; o sistema local de conhecimento. In: Petersen, P.; L. Silveira; E. Sabourin, *Agricultura familiar e Agroecologia no Semiárido; avanços a partir do agreste da Paraíba*. Rio de Janeiro, AS-PTA. p. 177-199.

Sabourin, E. (2009). *Camponeses do Brasil; entre a troca mercantile e a reciprocidade*. Porto Alegre, Garamond. (Col. Terra Mater).

Silva, Roberto Marinho Alves da (2006). Entre o combate à seca e a con- vivência com o Semi-Árido: transições paradigmáticas e sustentabilidade do desenvolvimento. Brasília – DF [Doctoralthesis – UNB]. http://repositorio.bce.unb.br/bitstream/Roberto/Marinho/Alves/da/Silva.pdf. Accessed January 2014

Silveira, L.; A. Freire; P. Diniz (2010). Polo da Borborema: ator contemporâneo das lutas camponesas pelo território. *Agriculturas*. Rio de Janeiro, AS-PTA. v. 7:1, p. 13–19.

Chapter 2

ANAFAE (n.d.). "Violaciones DDHH En Proyectos Extractivistas en Honduras." Al Consejo de Derechos Humanos de Nacionades Unidades 19 Sesión Grupo de Trabajo EPU, 2-15. Accessed December 6, 2016. https://drive.google.com/file/d/0B1ZA8HzEPi6jbENfM0VpZ2s1S0U/view

Boyer, Jefferson (2010). "Food security, food sovereignty, and local challenges for transnational agrarian movements: the Honduras case." *The Journal of Peasant Studies,* 37(2010):323-4.

Breslin, Patrick (2008). "The Agricultural Gospel of Elías Sánchez." *Grassroots Development* 29/1. Accessed November 7, 2016. http://thegoodgarden.org/pdf/ Don_Pedro.pdf.

Escoto, Edwin (2015). Internal Report to Groundswell International.

Espinoza, José Luis, Paola Sánchez, and Efraín Zelaya (2013). "Fincas agroecológicas en el bosque seco de Honduras." Asociación Nacional para el Fomento de la Agricultura Ecológica, October.

Frank, Dana (2013). "Hopeless in Honduras? The Election and the Future of Tegucigalpa." *Foreign Affairs*, November 22, 2013. Accessed November 7, 2016. https://www.foreignaffairs.com/articles/honduras/2013-11-22/hopeless-honduras.

Gao, George (2014). "5 facts about Honduras and immigration." Pew Research Center, August 11, 2014. Accessed November 7, 2016. http://www.pewresearch.org/fact-tank/2014/08/11/5-facts-about-honduras-and-immigration/.

Holt-Giménez, Eric (2001a). "Measuring Farmers' Agro-ecological Resistance to Hurricane Mitch in Central America." *International Institute for Environment and Development, IIED*. Gatekeeper Series No. SA102.

Holt-Giménez, Eric (2001b). "Midiendo la resistencia agroecológica contra el huracán Mitch." *Revista LEISA*, July v. 17:1, p. 7-10

Kerssen, Tanya (2013). *Grabbing Power: The New Struggles for Land, Food and Democracy in Northern Honduras,* 10. Oakland: Food First Books.

Nelson, Melissa (1998). "Hope Renewed in Honduras Mitch Teaches Lesson About Deforesting Land." *The Oklahoman*, December 16, 1998.

Smith, Katie. 1994. *The Human Farm: A Tale of Changing Lives and Changing Lands*. West Hartford, CT: Kumarian Press.

The World Bank (2011). "Nutrition at a Glance: Honduras." Document 77172, April 1. Accessed November 7, 2016. http://documents.worldbank.org/curated/en/617431468037498125/pdf/771720BRI0Box000honduras0April02011.pdf.

United Nations Office on Drugs and Crime (2013). "Global Study on Homicide." Vienna. Accessed November 7, 2016. http://www.unodc.org/documents/gsh/pdfs/2014_GLOBAL_HOMICIDE_BOOK_web.pdf.

WFP Honduras (2015). "WFP Honduras Brief." July 1-Sept. 30, 2015. Accessed November 7, 2016. http://documents.wfp.org/stellent/groups/public/documents/ep/wfp269059~6.pdf.

World Neighbors (2000). "Reasons for Resiliency: Toward a Sustainable Recovery after hurricane Mitch." *Lessons from the Field.*

Chapter 3

Conseils, Formation, Monitoring en Développement (CFM (2014). "Evaluation of PDL strategies for scaling agro-ecological farming alternatives." November 12.

Lentfer, Jennifer (2013). "USAID's answer to Oxfam on the article on the WINNER project in Haiti." *The Politics of Poverty*, October 17. Accessed November 7, 2016. http://politicsofpoverty.oxfamamerica.org/2013/10/usaids-answer-to-oxfam-on-the-article-on-the-winner-project-in-haiti/.

USAID (2012). "WINNER Main Achievements: Agricultural productivity increased." Accessed November 7, 2016. http://www.winnerhaiti.com/index.php/en/main-achievements/agricultural-productivity-increased

Chapter 4

Altieri, Miguel (2011). "Agroecología: Bases científicas para una agricultura sustentable." Montevideo: Nordan–Comunidad.

Anonymous (2014). "Convocatoria a II Congreso de Agroecologia Oct 2014." In *Biodiversidad en América Latina y El Caribe.*

Barrer, V., C. Tapia and C. Monteros C. (eds) (2004). *Raíces y Tubérculos Andinos: Alternativas para la conservación y uso sostenible en el Ecuador.* Instituto Nacional de Investigaciones Agropecuarias (INIAP), Quito.

Benzing, A. (2001). *Agricultura Orgánica. Fundamentos para la región Andina.* Villingen-Schwenningen, Germany: Neckar-Verlag.

Boada, L. (2013). "Prácticas alimentarias: relación con la diversidad en la alimentación en las familias campesinas de las comunidades: Ambuquí, Jesús del Gran Poder y Chitacaspi." MSc thesis, FLACSO, Quito.

Borja, R, S. Sherwood and P. Oyarzún (n.d.). "Katalysis: 'People-Centered Learning-Action Approach for Helping Rural Communities to Weather Climate Change. Informe Final de Sistematización'." CONDESAN-EkoRural.

Castro M.A. (2007). "La distribución de la riqueza en el Ecuador." In *Observatorio de la Economía Latinoamericana* 75.

Chiriboga, M. (2001). "Diagnóstico de la comercialización agropecuaria en Ecuador implicaciones para la pequeña economía campesina y propuesta para una agenda nacional de comercialización agropecuaria." Quito

Chiriboga, M. (2012). "Globalización y Regionalización: desafíos para la agricultura familiar ecuatoriana." RIMISP.

Daza, E. and M. Valverde (2013). "Avances, experiencias y métodos de valoración de la Agroecología. Estado del arte, mapeo de actores y análisis metodológico y de indicadores para la agroecología." IEE, Quito.

De Noni, G. (n.d.). "Breve vision histórica de la Erosión en el Ecuador." In *La Erosión en el Ecuador,* Centro Ecuatoriano de Investigación Geográfica, 44. Quito: ORSTROM.

De Schutter, Olivier (2010). "Report submitted by the Special Rapporteur on the right to food." United Nations Human Rights Council, 16th session, agenda item 3.

Fonte, S., S. Vanek, P. Oyarzun P, S. Parsa, D. Quintero, I. Rao and P. Lavelle (2012). "Pathways to Agroecological Intensification of Soil Fertility Management by Smallholder Farmers in the Andean Highlands." In *Advances in Agronomy,* edited by Donald L. Sparks, 125-184. Burlington: Academic Press.

Hidalgo F. et al (2011). "Atlas sobre la tenencia de la Tierra en el Ecuador," SIPAE.

IAASTD (2014). "Agriculture at a crossroads: Synthesis report. A Synthesis of the Global and Sub-Global." Washington, D.C., IAASTD Reports.

IFOAM (2011). "Position Paper: El papel de los campesinos en la agricultura orgánica." Germany.

INEC, MAG and SICA (2001). "Tercer Censo Agropecuario del Ecuador." Quito

La Vía Campesina (2011). "La Agricultura Campesina sostenible puede alimentar el mundo. Documento de Punto de Vista de la Via Campesina." Jakarta.

Marsh, K. (2011). "Una investigación en la práctica de Agroecología en Tzimbuto-Quincahuán." Internship report for EkoRural, Quito/ Universidad TREND, Canada.

Nwanze, K. (2011). "Viewpoint: Smallholders can feed the world." Rome, IFAD.

Oyarzún, P., R. Borja, S. Sherwood, and V. Parra (2013). "making sense of agrobiodiversity, diet, and intensification of smallholder family farming in the Highland Andes of Ecuador." *Ecology of Food and Nutrition* 52:515-541

Poinsot, Y. (2004). "Los gradientes altitudinales y de accesibilidad: dos claves de la organización geo-agronómica andina." *Cuadernos de Geografía* 13:5-20.

Poinsot, Y. (2011). "Los gradientes altitudinales y de accesibilidad." Op Cit

Tapia, C., E. Zambrano and A. Monteros (2012). *Estado de los Recursos Fitogenéticos para la Agricultura y Alimentación en el Ecuador,* Instituto Nacional de Investigaciones Agropecuarias (INIAP), Quito.

Zebrowski, C and B. Sánchez (1996). "Los costos de rehabilitación de los suelos volcánicos endurecidos." Report of the III international symposium about hardened vulcanic soils, Quito.

Chapter 5

Gliessman, S.R. (2015). *Agroecology: The Ecology of Sustainable Food Systems.* Boca Raton, FL: CRC Press/Taylor & Francis Group.

Gliessman, S.R., M.R. Werner, S. Swezey, E. Caswell, J. Cochran, and F. Rosado-May. (1996). "Conversion to organic strawberry management changes ecological processes." *California Agriculture* 50(1):24-31.

Koike, S., C. Bull, M. Bolda, and O. Daugovish (2012). "Organic Strawberry Production Manual." University of California Division of Agriculture and Natural Resources, Publication number 3531, Oakland, CA.

Monterey County Agricultural Commissioner (2013). "Monterey County Crop Report 2012." Salinas, CA.

Muramoto, J., S.R. Gliessman, S.T. Koike, C. Shennan, D. Schmida, R. Stephens, and S. Swezey. (2005). "Maintaining agroecosystem health in an organic strawberry/ vegetable rotation system." White Paper

Muramoto, J., S.R. Gliessman, S.T. Koike, C. Shennan, C.T. Bull, K. Klonsky, and S. Swezey. (2014). "Integrated Biological and Cultural Practices Can Reduce Crop Rotation Period in Organic Strawberries." *Agroecology and Sustainable Food Systems,* 38(5):603-631.

Sances, F., N. Toscano, L.F. LaPr, E.R. Oatman, M. W. Johnson. (1982). "Spider mites can reduce strawberry yields." *California Agriculture,* 36(1):14-16.

Santa Cruz County Agricultural Commissioner (2012). "Santa Cruz County Crop Report 2012." Watsonville, CA.

Shennan, C., J. Muramoto, S. Koike, M. Bolda, O. Daugovish, M. Mochizuki, E. Rosskopf, N. Kokalis-Burelle, and D. Butler. (2010). "Optimizing anaerobic soil disinfestation for strawberry production in California." *Proceedings of the Annual International Research Conference on Methyl Bromide Alternatives and Emissions Reductions,* 23.

Swezey, S.L., D.J. Nieto, J.R. Hagler, C.H. Pickett, J.A. Bryer, and S.A. Machtley. (2013). "Dispersion, Distribution, and Movement of Lygus spp. (Hemiptera: Miridae) in Trap-Cropped Organic Strawberries." *Environmental Entomology* 42(4): 770-778.

Chapter 6

Eijkennar, Jan (2015). "End of Mission Report: Resilience and AGIR." European Commission Directorate-General For Humanitarian Aid And Civil Protection – ECHO Regional Support Office For West Africa, 7 April.

Haub, Carl and Toshiko Kaneda (2014). "World Population Data Sheet." Population Reference Bureau.

IPCC (2008). "Synthesis Report: Contribution of Working Groups I, II and III to the Fourth Assessment Report of the Intergovernmental Panel on Climate Change." IPCC, Geneva.

IRIN. 2008. "Backgrounder on the Sahel, West Africa's poorest region." June. Accessed November 7, 2016. www.irinnews.org/report/78514/sahel-back- grounder-on-the-sahel.

Mathys, E., E. Murphy, M. Woldt (n.d.). "USAID Office of Food for Peace Food Security Desk Review for Mali, FY2015-FY2019." Washington, DC: FHI 360/FANTA.

Potts, M., E. Zulu, M. Wehner, F. Castillo, C. Henderson (2013). *Crisis in the Sahel: Possible Solutions and the Consequences of Inaction.* Berkeley: The Oasis Initiative.

Steyn, Anne-Marie (2015). "Opinion: To Solve Hunger, Start with Soil." *Inter Press Service News Agency*, April. Accessed November 7, 2016. www.ipsnews./net/2015/04/opinion-to-solve-hunger-start-with-soil.

Watt, Robert (2012). "Adopt or Adapt: The political economy of 'climate-smart agriculture' and technology adoption among smallholder farmers in Africa." CARE International SACC Project Report.

USAID (2014a). "Latest Sahel Fact Sheet." September. Accessed November 7, 2016 www.usaid.gov/crisis/sahel.

USAID (2014b). "Organizational Survey and Focus Groups on Adaptive Practices." November. Accessed November 7, 2016. http://community.eldis. org/.5c1fe9f0.

Chapter 7

Dembélé, Peter (2015). Internal Report to Groundswell International.

Diakité, Mamadou (1995). *Natural Regeneration Poem.* Excerpt translated by authors.

Gana, Drissa (2015). Internal Report to Groundswell International.

Gubbels, Peter, Ludovic Conditamde, Mamadou Diakite, Salifou Sow, Drissa Gana, Housseini Sacko (2011). "Raport Final: Auto-evaluation Assistee du projet "Trees for Change," Cercle de Bankass, Mali." January.

Gubbels, Peter, Bianovo Moukourou, Amadou Tankara, Oumar Sidibe, and Mama Traore (2013). "End of Project (2010-2013) Evaluation Report, Sahel Eco and International Tree Foundation, Assisted Self Evaluation of the "Re-greening Sokura" Project, Commune of Sokura, Mopti, Mali." November

République de Mali Institut National de la Statistique (2010). "Resultats Provisoires RGPH 2009."

Chapter 8

Batta, Fatoumata (2015). Internal Project Report to Groundswell International.

International Monetary Fund (IMF) (2012). "Burkina Faso: Strategy for Accelerated Growth and Sustainable Development 2011-2015." Country Report No.12/123, 10.

Jahan, Selim (2015). "Human Development Report 2015: Work for Human Development." United Nations Development Program, New York.

Chapter 9

CIKOD (2015). "The Hope is Now a Reality: FMNR on its Second Year," *FMNR Newsletter*, Vol 2, Issue 1, July.
Ghana Statistical Service (2015). "Demographic and Health Survey 2014." Accra, October, 155.
Guri, Bern (2015). Internal Report to Groundswell International.
Hjelm, Lisa and Wuni Dasori (2012). "Ghana Comprehensive Food Security and Vulnerability Analysis, 2012." World Food Programme, April.
Oppong-Ansah, Albert (2012). "Surviving on a Meal a Day in Ghana's Savannah Zone." Inter-Press Service, August 15. Accessed November 7, 2016, http://www.ipsnews.net/2012/08/ surviving-on-a-meal-a-day-in-ghanas-savannah-zone.
Toboyee, Juliana (2013). "Report on a Trip to FMNR Sites in Upper East Region, Bolga, Ghana." CIKOD, June 20-22.
Yussif, Mohammed Mudasir and Laminu Moshie-Dayan (2016). "Evaluation of the 'Eco-agriculture in Sahel Project in Ghana': A report of the findings from quantitative and qualitative fieldwork." March, 33.

Chapter 10

De Boer, H.C., M.A. Dolman, A.L. Gerritsen, J. Kros, M.P.W. Sonneveld, M. Stuiver, C.J.A.M. Termeer, T.V. Vellinga, W. de Vries & J. Bouma (2012). "Effecten van kringlooplandbouw op ecosysteemdiensten en milieuk-waliteit." Een integrale analyse op People, Planet & Profit, effecten op gebiedsniveau, en potentie voor zelfsturing, met de Noardlike Fryske Wâlden als inspirerend voorbeeld. Wageningen Livestock Research Report.
de Rooij, S. (2010). "Endogenous initiatives as driving forces of sustainable rural development." In *Endogenous development in Europe*, edited by S. de Rooij, P Milone, J. Tvrdonava and P. Keating, 29. Compas: Leusden.
Holster, H.C, M. van Opheusden, A.L. Gerritsen, H. Kieft, H. Kros, M. Plomp, F. Verhoeven, W. de Vries., E. van Essen, M.P.W. Sonneveld, A. Venekamp (2014). *Kringlooplandbouw in Noord-Nederland: Van marge naar mainstream*. Wageningen UR: Wageningen.
Noardlike Fryske Walden (2014). "Jaarverslag 2013." Burgum.
Sonneveld, M.P.W., J.F.F.P. Bos, J.J. Schröder, A. Bleeker, A. Hensen, A. Frumau, J. Roelsma, D.J. Brus, A.J. Schouten, J. Bloem, R. de Goede and J. Bouma (2009). "Effectiviteit van het Alternatieve Spoor in de Noordelijke Friese Wouden." Wageningen UR.
Stuiver, Marian (2008). "Regime Change and Storylines: A sociological analysis of manure practices in contemporary Dutch dairy farming." Wageningen: Wageningen University and Research Centre.
van der Ploeg, J.D. (2008). *The new peasantries: Struggles for autonomy and sustainability in an era of empire and globalisation*. London: Earthscan.
Verhoeven, F.P.M., J.W. Reijs, J.D. Van Der Ploeg (2003). "Re-balancing soil-plant-animal interactions: towards reduction of nitrogen losses." *NJAS Wageningen Journal of Life Sciences* 51(1-2):147-164.

Chapter 10

Gilbert, Natasha. 2012. "One-third of our greenhouse gas emissions come from agriculture." Nature, October 31.

Kerssen, Tanya (2015). "Food sovereignty and the quinoa boom: challenges to sustainable re-peasantisation in the southern Altiplano of Bolivia," *Journal of Peasant Studies,* 36:489-507.

Owen, James (2005). "Farming Claims Almost Half of Earth's Land, New Maps Show." *National Geographic News*, December 9.

United Nations (UN) (2014). "Open Working Group for Sustainable Development Goals." https://sustainabledevelopment.un.org/index.php?page=view&type=40 0&nr=1579&menu=1300.

UN Framework Convention on Climate Change (2015). Paris Climate Agreement, December 12. http://ec.europa.eu/clima/policies/international/negotiations/paris/index_en.htm.

Appendix 1

Uphoff, Norman, ed. (2002). *Agroecological Innovations: Increasing Food Production with Participatory Development.* London: Earthscan.

Index